MEAT

MEAT

A Natural Symbol

Nick Fiddes

London and New York

First published 1991
by Routledge
11 New Fetter Lane, London EC4P 4EE

Simultaneously published in the USA and Canada
by Routledge
a division of Routledge, Chapman and Hall, Inc.
29 West 35th Street, New York, NY 10001

Typeset in New Baskerville
by GilCoM Ltd, Mitcham, Surrey

Printed and bound in Great Britain by
Biddles Ltd. Guildford and King's Lynn

British Library Cataloguing in Publication Data
Fiddes, Nick, 1960–
Meat: a natural symbol
1. Food. Sociocultural aspects
I. Title
306

Library of Congress Cataloging in Publication Data
Fiddes, Nick, 1960–
Meat, a natural symbol/Nick Fiddes.
p. cm
Includes bibliographical references.
1. Meat. 2. Food – Symbolic aspects. 3. Food habits.
4. Food preferences. 5. Carnivora. 6. Meat industry and trade.
I. Title. GT2868.5.F53 1991
394. 1–dc20
90-23440
CIP

ISBN 0-415-04837-0

CONTENTS

FIGURES

PREFACE

As I recall, my interest in meat arose from a vague, but distinct, curiosity deriving from various episodes in my past. I have a vivid memory of a childhood holiday in France – I must have been about six at the time – when we children befriended a calf to which we gave the name Paddington, and my confusion on learning that we might well be eating Paddington before long, once he had grown up a little. I now know that the shock I felt is something that many children experience at some stage. I have been surprised at the numbers of parents encountered in the course of this study who related the story of their offspring's rebellion against meat, at whatever age. But, as with most children, I soon learned to accept the situation as normality.

I remember later conversations with my parents about why we call cows 'beef' and pigs 'pork' when they appear on the dinner plate – I must by now have been about 10 or 12. The precise explanation offered I do not now recall, except that it was somehow felt more polite to use these terms than their literal equivalents. This my youthful pride found difficult to accept: if we're going to eat cows and pigs, then why not be honest about it? Why try to dress it up as something else, as if we are afraid of facing the truth? I never did receive a satisfactory answer, and took a somewhat mischievous delight in occasionally referring to meat as 'dead animal' in situations where well-brought up children should not have done so.

As a teenager, the memory of a geography lesson comes to mind, when our teacher passed around a photocopied sheet about world food production, and the feeling of amazement that lingered for days on learning that there was no shortage of food in the world, but that the mass-starvation of which I had long been aware was largely the result of unequal distribution. I remembered this some years later in a geography lecture at university on learning about the ecological inefficiency of a large population eating 'high on the food chain': that it makes about ten times more sense in efficiency terms to eat the grain itself than to feed it to cows and then eat them.

Years later still, while I was running a catering business with a partner, one contract we received was for an academic workshop at Edinburgh University (Centre for Human Ecology 1985) on the subject of meat. The food we prepared was to complement the discussions, with leaflets accompanying each meal describing the dishes provided, and we kept in touch with the proceedings. The most striking thing I learned in the process was how *little* was known about the social aspects of the phenomenon of meat eating. Arguments ranged around nutritional, historical, economic, political, and environmental influences, but it was clear that none of these could sufficiently explain the centrality of meat in the conventional diet – yet no one seemed able to offer more than anecdotal evidence about *why* meat was so important in the first place.

But, in the end, I think the most significant influence that stimulated me to consider this enquiry was an even more personal one. For many years I had been aware that many of my friends were vegetarian, and yet I was not. I had heard many convincing arguments against meat eating, and yet I had never felt willing or able to give it up. Sometimes I felt that I should, but could not, and so felt a sort of guilt at my lack of will and moral failure. At other times I reasoned that it was perfectly natural and not unhealthy in moderation, so why worry? The result was internal struggle, and general unease. I wanted to know why this issue had the power to confuse me so much.

The work has grown out a doctoral thesis conducted over three years in the department of Social Anthropology at Edinburgh University (Fiddes 1989). My approach was to assemble and interpret as multifarious a range of examples as possible of the part that meat plays in our lives. Information was thus culled from academic material, from the popular media, and from individual conversations and a series of about fifty tape-recorded interviews. Many of these interviews were with individuals who have a special interest in meat, either for or against, but the quotes in this text are offered mainly to illustrate ideas and the sorts of way in which they are represented in everyday usage, and are not meant to represent particular people. Certainly I found no significant viewpoints peculiar to social groups, such as 'working-class' or young people. Rather, the ideas discussed are in general circulation, and are the stuff of constant interpretation, adaptation, and re-presentation throughout society.

What I have learned has forced me to reassess much about my entire life, not only in my attitude to meat but also in my approach to many other things: about the ways in which we behave towards each other and towards the habitat which sustains us. I feel that I have had to come face-to-face with an aspect of my identity of which I had previously been largely unaware, and that I might otherwise have continued to prefer not to recognise. Indeed, I now believe the very fact that we generally choose not to recognise certain important aspects of the meaning of our words and deeds to be significant in itself.

The analysis that follows uses a range of examples of the many ways in which we think about meat, talk about meat, and use meat – to look beyond the façade of the generally recognised, and to construct a coherent interpretation which accounts also for those ideas which are less often made explicit, but which may be effective nonetheless. It is not intended to imply that the symbolic notions associated with animal flesh provide a total explanation for the consumption patterns that exist. Clearly other factors such as nutrition are relevant. However, it is also clear that such influences, so often taken as objectively causal, are invariably subject to interpretation in contexts influenced by received ideas. This analysis is an attempt to redress the balance and, sometimes by deliberate overstatement, to demonstrate the importance of a social side of food habits which is all too often overlooked by a society convinced of its sophisticated rationality. This is an anthropological study of meat but it is also, from a particular angle, an ethnography of 'us': a society in the process of change.

My sincere thanks are due to everyone who has given me advice, assistance, motivation, and money. In particular I wish to thank Anna Ashmole for inspiring me to set out on this study, for her support and advice as it progressed, and for still being around to help so much at the end; Dr Alan Campbell and Dr Mary Noble of the department of Social Anthropology at Edinburgh University for being such excellent supervisors, and more; my parents, Brian and Dorothy Fiddes for their unstinting love and support without which I could never have got so far, and especially my mother for proof-reading; Ulrich and Francesca Loening at the Edinburgh University Centre for Human Ecology for bringing the topic to my attention; Professor Sir Keith Thomas for his astute criticisms and kind compliments; Miranda France for her perceptive eleventh-hour commentary; Geoff Harrington of the Meat & Livestock Commission for his generous contributions and stimulating ideas; all my colleagues and friends for their invaluable comments, cuttings, and encouragement; and not least to the many people who gave so kindly of their time and patience in indulging me with my tape recorder. I am also eternally grateful for the support of the Carnegie Trust; Fitch Lovell plc; the Konrad Zweig Trust; Greg Sams; the University of Edinburgh Vans Dunlop scholarship committee; and the Vegetarian Society.

Nick Fiddes
1991

INTRODUCTION

This book is about meat. It is about what parts of which animals we habitually eat, when we eat meat, where we eat it, and with whom we eat it. Principally, it is about why we eat it, or why we do not. The central aims are to suggest, firstly, how meat has come to be the special food it is in Britain today – not why we eat meat at all, but why we eat so much, so often, and with such ceremony; and, secondly, to suggest why certain changes in our eating habits may be happening now.

The answers to these questions may seem so obvious as to be scarcely worth mentioning. It is easily taken for granted that meat is an important part of the diet because it is high in strength-giving protein, and simply because when cooked it tastes good and is satisfying. The fact of changes in eating habits are likewise routinely explained as fashion, or by reference to worries about high levels of saturated fats, or chemical residues, or perhaps about the cruelty involved in intensive husbandry . . . 'Where is the problem?', one might ask.

Such common-sense beliefs must be questioned, since what seems natural fact to us, in our particular society, at this particular time, is exposed as mere cultural orthodoxy when set against the range of beliefs and practices of other societies and in history. Many people live healthily with little or no meat in their diets. Others subsist almost exclusively upon it. And, as the meat industry is quick to point out, the health fears about eating meat seem, sometimes at least, to be out of proportion to the real physical threat involved.

Why are health concerns about the allegedly high fat content of meat being expressed now? In the Punjab 'You look fat' is said to be a compliment (Naipaul 1964: 80) and elsewhere meat might be believed to make one thin. Even today many people still hold to the notion (wrongly, we are now told) that it is potatoes or rice that are most fattening. Why, whilst becoming less expensive as a proportion of average income, have purchases of the traditionally prestigious red meats been falling? And why have ethical concerns recently come to prominence, when for years most people have been happy to consume animal products without such worries? Our conventional explanations are not entirely adequate. Fuller answers must be sought by interpretation of what meat stands for in our culture.

The title of this work suggests that meat is a 'Natural Symbol'. This plays on Mary Douglas's work about bodily symbolism entitled *Natural Symbols* (1970), in which she showed how the human body is an immediately accessible and therefore natural metaphor for the expression of social experience. In similar vein, the global occurrence of certain ideas suggests that they may tend to arise by the very 'nature' of meat. This should not be surprising. We know that, biologically, food selection and consumption are highly significant – Young, for example, argues that 'food is about the most important influence in determining the organization of the brain and the behaviour that the brain dictates' (1968: 21). We know that societies use systems of classification to regulate their internal and external relations (e.g. Douglas 1973). This work suggests that our use of meat as a food reflects our categorisation of, and our relations towards, animal competitors, companions, and resources. Perhaps then it is only 'natural' that meat should be so widely selected for special social or ritual significance, even if only by its avoidance.

Calling meat a 'Natural Symbol' also, however, refers to the central organising idea of the work. The analysis is centred upon the argument that the most important feature of meat – which endows it with both its positive image as prestigious and vital nutrition, and simultaneously its contrary image as dangerously immoral and potentially unhealthy – is that it tangibly represents human control of the natural world. Consuming the muscle flesh of other highly evolved animals is a potent statement of our supreme power.

At this point it must be emphasised that this does not necessarily mean that we each consciously glory in the subjugation of animals whenever we bite into a piece of meat. The associations with environmental control which underpin meat's generally high status in our communal system of values do not have to be held consciously, individually, to be effective. It is rather that the principle of power over nature is an omnipresent thread running through the culture in which we are raised, and which forms the context of our thought and debate. It is not an invisible thread, but we usually do not see it, for reasons that will be explored. Through much of British history, and western history in general, human subjugation of the 'wild' natural world has been a central theme – at times an almost religious imperative – and I shall show that consumption of animal flesh has been an ideal exemplar of that control. Despite our rationalisations and refinements, modern scientific civilisation is no exception to this; meat still derives its peculiar significance from these basic ideas. In this sense too, meat is a Natural Symbol.

We define meat as the flesh of animals destined for our consumption. Until about the fourteenth century, 'meat' could mean any nourishment. Over time, usage of the term reduced in scope to mean animal food (much as 'drink' has begun particularly to denote alcohol); more recently its meaning has tended to exclude even poultry or fish. According to the the book of Genesis (i. 28) meat is potentially derived from 'every living thing' that moves, although we classify many creatures, including those of our own species, out of normal consideration. I do not intend to define meat any more closely than this. To do so would only invite unnecessary definitional dilemmas such as those which confront researchers into vegetarianism who have found it necessary to distinguish, firstly, those who avoid only red meat, from those who avoid also poultry, from those who avoid also fish, dairy produce, animal products, and so on (e.g. Dwyer *et al.* 1973). Fine distinctions, and a scientistic terminology of ovolactovegetarians and pescovegetarians and the like, are doubtless vital in some contexts, but are not the concern here. I have met one man who will eat red meat but not fish, another who will eat poultry but avoids dairy products, and another concerned about the classification of the yeast in his daily bread. Such permutations of habit cannot be, and need not be, conveniently reconciled within a single neat categorisation.

Meat, instead, is taken to mean simply that which people regard as meat. If one person thinks only beef to be meat then that, for them, is what meat is. If another includes also lamb, poultry, game and fish then so too, for that person, that is the definition of meat. Thus, on the whole, in the context of Britain and most Western societies, the word applies most commonly to so-called red meats – the flesh of domesticated cows and perhaps sheep – and also to pork and to game. Poultry, and especially fish, is rather less 'meat' to many people (some reasons for which will be considered later) though may still be included. The subject under scrutiny is not the substance, but the concept. Meat, for our purposes, is just what you, and I, and the man on the Clapham omnibus, refer to as meat.

Whilst this is not a study of vegetarianism, references to it will be found throughout the text. This is because meat eating and vegetarianism are two sides of the same coin – each being significant in opposition to the other. Research into vegetarianism, such as the above example, generally encounters a problem of definition: how to classify the variety of beliefs and motivations that are offered as explanations for that inclination. This commonly leaves writers baffled for lack of a uniting factor. The error is twofold: firstly, in expecting the term 'vegetarianism' to have a single definitive characteristic, rather than a range of possible features: which Needham refers to as a polythetic array of serial and more complex resemblances (1983: 36-65). But secondly, and more importantly, the problem is in looking for the nature of the preference within vegetarianism itself, when its definition, in the end, lies not in what it is, but in what it is not.

Vegetarians do not eat meat (or, at least, some meats). Although it is often overlooked, the one and only attribute which characterises all vegetarians, regardless of race, creed, class, gender, age, or occupation, is an avoidance of animal flesh in their diet. Thus, transparently, the question of what motivates vegetarians can only be adequately answered by considering what motivates meat eaters – what it is about meat the makes people want to eat it – since rejection of such beliefs is the one thing that vegetarians have in common. (I exclude those who eat meat gladly but rarely because of its expense.)

The absence of more than superficial consideration of the reasons for meat eating in much of the literature on vegetarianism may lie partly in the conventional assumption of the majority of the population that meat is a normal, natural part of the diet, and vegetarianism an aberration to be explained.

Indeed, in British society, until recently, that has broadly been the case. Children have traditionally been brought up to regard consuming the flesh of other animals for food as both normal and desirable. Meat eating is part of what Bourdieu (1977) calls our 'habitus' – it is a principle unquestioned by most people. That this traditional view is implicit in much published research is obvious from the language commonly used: of 'faddism', 'rebelliousness', or 'deviance'. It would be easy to find any number of people who would agree that vegetarianism is generally ideological, if not overtly political. It would be harder to persuade most of those same people that meat eating is likewise. Nonetheless, any study of food habits must recognise that food selection is imbued with social rules and meaning, and it is clear from the extent of its association with cultural rituals, both religious and secular, that meat is a medium particularly rich in social meaning. From an academic viewpoint, therefore, a prejudice in favour of the majority is unsatisfactory. All that can be said is that food habits differ, and the meat-eating habit requires explanation as much as does the non-meat-eating habit.

The analysis of notions of meat may help to explain why increasing numbers of people have been avoiding meat in their diets. Conversely, study of the beliefs of vegetarians is a prerequisite to understanding the phenomenon of meat consumption. With a habit such as meat eating, which has traditionally been so taken for granted as to be seen as the natural order, the ideas which underpin the belief can be hard to elucidate. However, by bringing into consideration also the ideas of those who reject the tenets of meat eating, it is made more possible, by opposition, also to illuminate some of the distinctive features of that set of beliefs and values. Our attitudes to meat, I suggest, are a reflection of our world view, and changing habits in meat consumption may well indicate a changing perception of the world we inhabit.

The book is arranged in four main parts, which are best approached in the order written. The first part provides necessary **Background** to the topic, although some readers may prefer to jump immediately to Part II. Chapter 1 illustrates the curious way in which, time and again around the world, meat is a particularly valued food, sometimes to the point of being the only 'real' food; and, in Chapter 2, some changes in meat consumption habits which have occurred over the years are set out. In Chapter 3 it is argued that the notion of 'taste' reflects – rather than explains – preferences. And Chapter 4 elaborates on the notion that meat, like all food, feeds not only our bodies, but also our minds; it is more than just a meal, it is part of our way of life.

The second part, **Meat is Muscle,** presents and expands the main argument of the analysis: that the high value of meat is largely contingent upon its symbolic importance as a tangible representation of human control of, and superiority over, nature. The fifth chapter looks at how we habitually relate the origins of the human species, and of human civilisation, to the advent of hunting and of farming respectively. Chapter 6 investigates the history of affirmations of human supremacy, and the importance of blood as a symbol of that supremacy, and then demonstrates the extent to which these values permeate each stage of the meat production and consumption system. In Chapter 7 an alternative ethos is described, in which humanity is conceived of as complementary to nature, rather than opposed to it; it is shown that this rival viewpoint has also influenced the meat system through much of Western history, and is expressed in our growing repugnance to reminders of meat's animal origins.

Mixed Meataphors, the third main part of the work, deals with some aspects of the meat system whose significance we seldom recognise. It shows how the symbolic importance of meat as an expression of environmental control accounts for peculiar details of the British food system that might otherwise seem obscure. Arens's (1979) contention that no adequate evidence exists for the occurrence of customary cannibalism, anywhere, is explored in Chapter 8, and the taboo is shown to conform to the traditional western orthodoxy that anything non-human is 'fair game' (unless proscribed for other reasons). This principle is then extended in Chapter 9 to explain our reluctance to eat pets, or animals that are otherwise classified as close to humans. In Chapter 10 it is suggested why meat should figure so regularly in sexual imagery in the English language.

Whereas the third part deals with aspects of the meat system which are more meaningful than we commonly realise, the fourth and final main part of the work deals with the standard explanations for the status of meat – our **Modern Meatologies** – and shows them each to have important symbolic aspects in addition to their overt and obvious meanings. Economics is one of the most pervasive influences of our age, but in Chapter 11 it is shown that economic techniques can only quantify the high value we place upon meat. The sources of that esteem must be sought elsewhere, and these are again related to meat's allusive function. Chapter 12 suggests that our health is likewise not the straightforward causative process of nutrition and contagion that we often assume; perceptions of the healthiness of meat express wider concerns about our relationship with the world that sustains us. In Chapter 13 some ethical and religious views of meat are discussed, and again are shown to reflect power relationships. Finally, Chapter 14 looks at some of the many ways in which meat production has recently been indicted as ecologically damaging – an involvement that is both literal and metaphorical. The concluding chapter sums up the principal findings, and speculates about possible future trends in the meat system.

Part I
BACKGROUND

1

FOOD = MEAT

Meat is a curious thing . . . In Uganda, plantain that would feed a family for four days exchanges for one 'scrawny' chicken with less than a twentieth of the nutritional value (Bennet 1954: 32). The Sharanahua of Peru see hunting for meat as men's primary occupation yet even the most active hunt for only a few hours on less than half the available days (Siskind 1973: 93). At Toraja funeral ceremonies in Indonesia exchange and division of meat makes important statements about status and themes of honour (Wellenkamp 1984). And we are told that the war-time German government's wish to supply its forces with 'excessive' standards of protein intake may have led necessarily to:

> a distortion of agriculture towards animal production and hence to a lower total food production and the country's inability to withstand the Allied blockade. In the United Kingdom, by contrast, the philosophy of the minimal diet appealed more and food supplies were preserved by a switch towards cereal production.
>
> (Rivers 1981: 20)

Whether or not Germany's military potency was indeed reduced, it is notable that its leaders apparently elected to supply their population with ample meat (or 'protein') at all cost. America, meanwhile, rationed meat for its civilians, although 'richly fatted beef was force-fed into every putative American warrior' (Baker 1973: 43). Around the world, meat plays a singular part in people's affairs, not least through a common association with strength and aggression.

Chagnon begins his description of the diet of the Yanomamo –

the 'Fierce People' – of Amazonia by noting that 'The jungle provides numerous varieties of food, both animal and vegetable. The most commonly taken includes several species of monkeys, two varieties of wild "turkey", two species of wild pig . . .' He then reveals that 'Game animals are not abundant, and an area is rapidly hunted out' (Chagnon 1977: 29, 33). In fact the Yanomamo spend almost as much time hunting as gardening although cultivated foods provide 85 per cent or more of their diet. Is it not rather odd that the Yanomamo allocate so much time to obtaining meat, and that Chagnon should devote the bulk of his attention to a food that constitutes less than 15 per cent of his subjects' diet?

In similar vein a modern textbook description of nutritional theory is headed by the 'proteins we need' which firstly 'come mainly from meat, fish, cheese, milk and eggs' (Matthews and Wells 1982: 1), having opened with a historical review which begins: 'In ancient times man was . . . continually on the move, living in tribes, getting his food by fishing or hunting for wild animals and foraging for edible plants and fruits and berries' (1982: i). The mention of 'man's' obtaining animal foods before (woman's?) foraging activity – although the latter may well have provided the bulk of the diet as in most modern subsistence societies – is significant for the very orthodoxy of its priorities. Nutrition in 'under-privileged countries' is discussed as a final chapter which blames ignorance and 'taboos' for preventing the 'best use' from being made of available food – especially for pregnant women who are said to 'have the highest protein requirements of the community and need all the animal protein foods to meet their increased needs' (1982: 232). The book advocates 'better education' by western agencies, with no recognition that our own exalted valuation of 'protein' might in fact *induce* deprivation in 'The hungry nations' since cash crops for the rich tastes of the affluent repeatedly rob traditional communities of their fertile land and labour (George 1984).

The primacy of animal protein has been an established tenet of nutritional wisdom for many years amongst 'experts' as well as amongst much of the public. In one recent study of food distribution within British families, when asked what the family needs to eat properly:

> meat was mentioned by the women more frequently than any other food. In fact, only five women [out of 200] thought meat was not an important item of the family diet. Meat, or fish as its substitute, was usually viewed as an essential ingredient of the main meal of the day and a proper meal was most commonly defined as meat and two veg . . . men's preference for meat ensured its regular consumption in most families, and when inflation or lowered income had an impact on family eating it was the reduction in the quantity and quality of meat which was most frequently reported and most regretted.
>
> (Kerr and Charles 1986: 140)

Time and again, in different contexts, cultures, social groups, and periods of history, meat is supreme. Within most nations today, the higher the income bracket, the greater the proportion of animal products in the diet. In one study of over 50 countries, higher-income groups consistently derived far more of their fats, proteins and calories from animal sources than did lower-income groups (Perisse *et al.* 1969). As Julia Twigg notes:

> Meat is the most highly prized of food. It is the centre around which a meal is arranged. It stands in a sense for the very idea of food itself . . . our meat and drink. At the top of the hierarchy, then, we find meat, and in particular red meat, for the status and meaning of meat is quintessentially found in red meat. Lower in status are the 'bloodless' meats – chicken and fish – and below these are the animal products – eggs and cheese. These are sufficiently high in the hierarchy to support a meal's being formed around them, though they are confined to the low status events – the omelette and cheese flan of light lunch or supper. Below these we have the vegetables, regarded in the dominant scheme as insufficient for the formation of a meal, and merely ancillary.
>
> (Twigg 1983: 21–22)

Meat is so significant that, all over the world, people describe a 'meat-hunger' that is unlike ordinary hunger. Among the Canela of Amazonia, for example, *ii mo plam* means 'I am hungry', whilst *iiyate* translates as 'I am hungry for meat' (Gross 1975: 532). Marvin Harris (1986: 31) correspondingly asserts that humans are genetically programmed to prefer animal foods. But it is the fact that Harris argues this case with conviction that is significant, since science does not otherwise support his view. 'Instinct' is more likely to be a useful *topic* than a useful *tool* for analysis. Both biological and anthropological evidence suggest that 'humans are food generalists . . . As a direct consequence of this, the recognition of foods cannot be pre-specified genetically' (Rozin 1976: 286). A food habit is rather 'a feature of society and is integrated into a structure of social values that may have nothing to do with the principles of nutrition' (Le Gros Clark 1968: 69). Like Harris, however, many of us believe in some biological 'instinct' – including this habitually meat-eating woman who, in an interview with me, described missing it whilst living amongst vegetarians:

> 'Well, on a subliminal level, I think, you just have this notion that it's going to be more filling; it's going to be nicer; you're going to have a better feeling of . . . you're not going to want, sort of, six cream cakes after it. There's really no doubt about it, that while we were eating vegetarian food we were always hungry. I mean, we always were . . . The fact was, that without some meat at all – I mean even if it was once every other day . . . even if it was a glimmer of meat – that without any meat at any time we always had a slight hunger.'

This sort of 'meat hunger' is widely expressed in a variety of ways. Meat is, to many, almost synonymous with 'real' food. To the habitual meat eater, such as a male marketing executive being proposed a vegetarian alternative by his wife, it can be difficult to imagine its absence: only meat has the right substance; only meat is proper food: 'No, I mean you can't sort of chew that. What do you chew on? What do you eat?'

Not only preferred food, meat is regularly synonymous with food – like bread, which enjoys a similar, though humbler, symbolic role in British tradition. The !Kung of the Kalahari, for example, describe gathered provisions as 'things comparable to nothing' while meat provided by men is called 'food' (Lee 1972; Shostak 1983). To be deprived of meat can indeed be equated with starvation. A middle-aged meat-eating British woman, for example, when asked how she might feel had she to kill animals for their meat herself, responds: 'I don't think I could. I think I'd probably starve.'

A similar confusion of 'food' with 'meat' emerges in academic writings. Lévi-Strauss, for example, is known for his structural analyses of food systems, based on practices and mythologies. In a series of publications (e.g. Lévi-Strauss 1963, 1966, 1970, 1973, 1978, 1987) he purports to demonstrate, amongst other things, how fire universally transforms food from a natural state to a cultural state demarcating, he argues, the emergence of humanity. By his later work the scheme has become highly complex, including a variety of cooking operations whereby he maintains food is alternatively naturally or culturally transformed, summed up in his celebrated 'culinary triangle' (1966). Lévi-Strauss's work is significant here for one point made, and one point missed.

The interesting point he makes is the importance of routine cooking as a human universal, at least on a par with language, distinguishing human civilisation from the rest of the natural world. The point missed is that Lévi-Strauss largely fails to acknowledge that in most cases he is not discussing the cooking just of *food*, but particularly the cooking of *animals*. When he argues, for example, that smoking is a 'natural' means of transforming food to 'cultural' ends he surely does not mean the smoking of parsnips or plantains, but of meat. Likewise, if roasting has a special status, as he maintains, it is the roasting of meat. Only in his most complex elaborations does Lévi-Strauss distinguish animal from vegetable by which time the point is well lost (1973). He too evidently regards meat as the quintessential food.

Lévi-Strauss is not alone in this. Seeger, for example, writing of the Suya people of Central Brazil, explains that their strongest expression for an odour – *ku-kumeni*, that might be translated as 'gamy' – refers to sexual excretions and slightly tainted, but not rotten, food (1981: 93). He clearly means tainted meat. Edmund Leach too, in his analysis of 'Animal categories and verbal abuse' (1964) persistently talks about 'food values' and 'food names' whilst his discussion revolves around animals and flesh foods to a degree utterly disproportionate to their role in the diet, measured by monetary value, nutritional value, or bulk. His habitual equation of 'food' with 'meat' merely reflects the preeminence that meat has in the western diet and thought.

Meat regularly takes the starring role. Caricatured though 'meat and two veg' may be, the inevitable reply to 'What's for dinner?' will be 'pork', or 'chicken', or 'beef', or whatever the meat component of the meal may be. Likewise, when dinner commences the brussels sprouts are unlikely to receive first comment:

> 'A lot of the things I enjoyed had nothing to do with meat. So I suddenly realised that, you know, most people always make this big thing about a meal, that it's the meat that's always most praised when someone cooks a meal and that's the important bit. And I enjoyed the other bits so much.'

The primacy of meat surely also underpins a seemingly insignificant, but slightly perplexing, minor personal routine:

> 'I know it doesn't make much sense but I always have to put the meat on the plate first before the vegetables. I really don't know why. It just seems right that way – meat first, vegetables next. I keep wondering why it is I do it but . . . it just wouldn't be right otherwise.'

The arrival of a 'roast' at table can be a scene of considerable ceremony. It is the one occasion in the traditional British household where the male head of household may be expected to help serve, as he may have supervised its purchase. The meat's arrival is properly greeted by conspicuous inhalations and references to its aroma, and the first mouthful should be followed by appropriate remarks on its flavour and tenderness. But whilst a roast of meat is still the epitome of the proper meal, it is the idea of any meat, the feeling of meat, the spirit of meat, that is essential:

> 'I do have meat with most meals I suppose. We might just have an omelette now and again, but usually there's at least a bit of meat there, like in a spaghetti bolognese or something. I mean, even if there's lots of vegetables and things there, it wouldn't taste the same without that bit of mince.'

The range of soya-based meat-analogues and other substitutes available today testifies to the centrality of the *concept* of meat, not to its dispensability. Many people wishing to avoid meat feel that the gap left in their habitual food system needs to be filled with a direct equivalent which mimics the form or the nutritional content of meat itself. At the launch of Quorn, a new 'high protein, fibrous substance brewed entirely from a microscopic plant', Saffron Davies asks:

> who will buy it? Vegetarians are an obvious target if they want to eat 'meat' that is not meat. It can be made to look more or less like herbivorous flesh, it chews like meat and it has a similar texture.
>
> (Davies 1988: 34)

But it seems likely that even if a perfect substitute for meat were developed, indistinguishable in any respect from the real thing, many meat eaters would be reluctant to swap. There is just something important about its having come from an animal. As the technical director of a company producing soya protein remarks, explaining the fact that more of his product is fed to pets than to humans, 'You do not have to educate dogs, except by giving them the stuff' (Clayton 1978: 6). Unfortunately, however, few meat substitutes are entirely satisfactory. With Tivall, for example, a recent soya-based contender in the field:

> making the bean palatable has been a major problem. Technology has made it possible to isolate the protein in the bean for use as Textured Vegetable Protein (TVP) or to mix with cereals in meat substitutes. But TVP is like trying to digest a minced trampoline, and all of the meat substitutes made in the United Kingdom have tasted so dire to me that I would prefer to go hungry. These products, trying to ape the British sausage, add too much cereal with its carbohydrate content while allowing a pronounced aftertaste to linger on the palate like a fermenting sock.
>
> (Spencer 1988: 23)

Even when the form of meat is entirely foregone, a substitute, such as cheese or eggs, is almost always of animal origin – possibly due to lingering belief in the need for large amounts of protein in a healthy diet. Meat and animal products are pre-eminent in our food system and, even allowing for the fact that the majority of ethnographies are written by western anthropologists with western interests, it is clear that this is also true of the food systems of many other cultures.

But meat is not only the most privileged nourishment; it is also the most feared and abhorred. The likeliest potential foods to nauseate us today are those recognisably animal – the gristle, the blood vessels, the organs, the eyes – unlike vegetable foods whose identity we rarely dread. Around the world, meat is by far the most common focus for food avoidance, taboos, and special regulation (Simoons 1967). Hindus, for example, revere cows and would not contemplate their consumption, whilst Jews and Moslems abhor pork as unclean. And in western society too, feelings of disgust about foods almost always relate to meat or other animal products (Angyal 1941). In Macbeth, for example, almost every component of the infamous witches' brew is of animal origin (IV. i):

FIRST WITCH

Round about the cauldron go;
In the poisoned entrails throw:
Toad that under cold stone
Days and nights has thirty-one.
Sweltered venom, sleeping got,
Boil thou first i'the charmèd pot.

ALL

Double, double, toil and trouble;
Fire burn, and cauldron bubble.

SECOND WITCH

Fillet of a fenny snake
In the cauldron boil and bake;
Eye of newt, and toe of frog,
Wool of bat, and tongue of dog,
Adder's fork, and blind-worm's sting,
Lizard's leg and howlet's wing,
For a charm of powerful trouble,
Like a hell-broth, boil and bubble.

ALL

Double, double, toil and trouble;
Fire burn, and cauldron bubble.

THIRD WITCH

Scale of dragon, tooth of wolf,
Witch's mummy, maw and gulf
Of the ravined salt sea shark,
Root of hemlock digged i'the dark,

Liver of blaspheming Jew,
Gall of goat, and slips of yew
Slivered in the moon's eclipse,
Nose of Turk, and Tartar's lips,
Finger of birth-strangled babe,
Ditch-delivered by a drab,
Make the gruel thick and slab.
Add thereto a tiger's chaudron
For the ingredience of our cauldron.

ALL

Double, double, toil and trouble;
Fire burn, and cauldron bubble.

SECOND WITCH

Cool it with a baboon's blood;
Then the charm is firm and good.

But in spite of all our reservations – and efforts to avoid its unsavoury associations – something uniquely important about meat's substance remains for most of us. However much we may feel uneasy about its source, those who feel able to abstain altogether remain a small, though perhaps growing, minority. Meat has the potential both to comfort and to nourish us. It can also confuse and confound us. It revolts and satisfies us. It meets a need in us. There is something very curious about meat.

2

A BRIEF HISTORY OF MEAT EATING

The advent of hunting – variously estimated at between two and four million years ago – marks the emergence of humanity, and prehistoric life revolved around securing meat . . . or so it is said. Jane Renfrew, for example, writes that 'The first men appear to have arrived in Britain sometime before 300,000 years ago. These men were hunters' (Renfrew 1985a: 6). This canon will be shown in Chapter 5 to be based largely on modern supposition. Indeed Renfrew tacitly admits that the common icon of early humankind's carnivorism is substantially conjectural: 'From the camp sites so far excavated, there has been little evidence for the plant food part of their diet, but on analogy with modern hunting communities up to 80 per cent of their diet may have consisted of vegetable sources' (1985a: 6).

Archaeological excavations do suggest that people have eaten some meat for as long as they have inhabited the British Isles – rings of mutton-bones found around fire-pits belonging to the Celtic Belgae tribe, for example, imply members throwing gnawed bones over their shoulders (Pullar 1970: 45) – but such evidence cannot accurately indicate *how much* meat was eaten, nor *how regularly* it was consumed. The archaeology of early Christian Ireland suggests that livestock husbandry was primarily based upon dairy farming rather than meat production (McCormick 1987) which is consistent with some evidence from later periods. But little is well established historically before about the time of the Norman invasion, and what information exists relates mainly to ruling élites. Surveying eating habits in Roman times, for example, Renfrew writes that:

> Perhaps the best introduction to Roman cooking is to look at the description of some of the most elaborate banquets recorded – bearing in mind that they are exceptions rather than the rule, they give one a vivid insight into the extravagant aspirations and achievements of the Roman cooks.
>
> (Renfrew 1985b: 11)

She catalogues dormice seasoned with poppy seeds and honey; 'eggs' made from spiced garden warblers in pastry; beef kidneys and testicles; the uterus of a sow; chickens; hare; wild boar containing live thrushes; pigs, slaughtered on the spot, stuffed with black pudding and sausages – and all at a single meal! This indeed tells us much about the excesses of a ruling class but little about how most people actually subsisted. After another dozen similar pages Renfrew reveals that 'the Romans were enthusiastic about vegetables' and provides an inventory of their delights in a paragraph of seven lines (1985b: 23) highlighting both the paucity of reliable data on the normal diet of pre-modern times and the bias of most modern writers towards the colourful lives of a minority.

From what evidence exists it can be gleaned that until as recently as the last few centuries animal products were for most people probably less pre-eminent than they are today. The Reverend Oswald Cockayne's studies of Anglo-Saxon manuscripts (1864), for example, clearly show animal dishes to have been just one part of the cook's repertoire; goose-giblets, pigs-trotters, and pigeon in a piquant sauce were lauded, but equally were peas with honey, and nettles cooked in water. Philippa Pullar, in her history of English food, holds that when much is made of the 'poverty' of diet in the period the fact that cattle were not then reared primarily for meat is not usually taken sufficiently into account. Oxen, she says, 'were draught animals, cows were for milk; sheep were for wool and dairy produce. The diet was largely one of dairy produce, legumes, cereals, game, fish, wild fowl and young animals' – though Pullar herself nonetheless illustrates this passage solely with meat dishes: 'Meat broths and stews containing pot-herbs were concocted in giant cauldrons; meat was also fried, steamed or roasted and brought to the table on long spits' (1970: 74).

Meat's proscription during Christian periods of fasting such as

Lent suggests that it was then already a prestige food. Until the eighth or ninth centuries this could mean as many as fifty or sixty successive days without meat and rules of abstinence could embrace half of all days in the year. Throughout the Middle Ages the greatest differences in eating patterns were not so much between geographical areas as between the mass of the population and a numerically small but outstandingly wealthy élite whose diet was marked by conspicuous consumption in terms of quality, quantity, and variety (Pullar 1970: 75; Kisbán 1986: 3–4). As Norbert Elias notes of the period:

> The relation to meat-eating moves in the medieval world between the following poles. On the one hand, in the secular upper class the consumption of meat is extraordinarily high, compared to the standard of our own times. A tendency prevails to devour quantities of meat that to us seem fantastic. On the other hand, in the monasteries an ascetic abstention from all meat-eating largely prevails, an abstention resulting from self-denial, not from shortage, and often accompanied by a radical depreciation or restriction of eating. From these circles come expressions of strong aversion to the 'gluttony' among the upper-class laymen. The meat consumption of the lowest class, the peasants, is also frequently extremely limited – not from a spiritual need, a voluntary renunciation with regard to God and the next world, but from shortage. Cattle are expensive and therefore destined, for a long period, essentially for the rulers' tables.
>
> (Elias 1939: 118)

It is the habits of the élite which are better documented and which tend therefore to characterise the period. Thus when we hear that medieval Europeans were exceptionally carnivorous compared to the vegetable-eating peoples of the East (Braudel 1974: 248–9) the comparison is with the wealthy and powerful. Of them the popular image is at least partly true with occasional lavish and ostentatious feasting. Otherwise meat was in generally short supply, except immediately after the Black Death when a smaller population had more and better land and stock to share.

The late Middle Ages saw a further stimulus to meat consumption with increasing use of draught horses gradually releasing oxen for human food (Thirsk 1978). By the eighteenth century England had more domestic beasts per acre and per person than any country in Europe except the Netherlands (O'Brien 1977: 169) and a longstanding reputation for meat consumption. A foreign visitor to England in the 1790s reported (presumably referring to the upper classes):

> I have always heard that they [the English] were great flesh-eaters, and I found it true. I have known people in England that never eat any bread, and universally they eat very little. They nibble a few crumbs, while they chew meat by whole mouthfuls.
>
> (quoted in Stead 1985: 20)

Another cause of increasing meat consumption in the eighteenth century was a series of agricultural innovations. New animal feeding practices and the enclosure of land removed the need for the slaughter of animals for salting prior to the onset of winter. Meanwhile, the import of new breeds from Holland markedly raised productivity. For many people meat from farm animals began to replace game meat for the first time, particularly as hunting laws became more restrictive for non-landowners. High meat consumption became general amongst more than just a powerful minority. 'Butchers meat was cheap' although 'if one compares the prices with wages it may be seen that working men could not afford to eat well' (Stead 1985: 23).

Technical innovations facilitated the increase in average meat consumption which is evident from the eighteenth century onwards. (All such figures must be treated with some caution since regular collection of agricultural statistics began in Ireland only in 1847 and in Great Britain in 1867 and UK meat output was not officially measured until 1907 (Perren 1978: 2).) But, as we shall see, it is important also to note that society's perception of – and thus relationship with – the world that it inhabited was undergoing substantial modification during this period. Rapidly changing scientific orthodoxy and expanding industrial potency were combining to alter the very way in which people viewed their surroundings as, to an unprecedented extent, people came to extol the virtues of environmental conquest. In the words of Eszter Kisbán, this period of history:

> embraces the emergence of modern natural sciences, technical innovations, industrialisation, urbanisation. Though

> they appeared at different periods in different places, they nevertheless provide the characteristic features . . .
>
> It is no accident that in parallel with these great economic and social changes there was a continuous increase in meat consumption.
>
> (Kisbán 1986: 8)

After a period of decline in the 'hungry forties', British meat consumption increased steadily from the mid-nineteenth century. By 1881 more money was spent on meat than on bread (Burnett 1966: 13, 129).

Reay Tannahill sees rising consumption as partly supply-driven since the latter decades of the nineteenth century, 'the heyday of imperialism, were years of land-grabbing and utilization on a majestic scale' (Tannahill 1988: 316). Perren suggests a general rise in real incomes, a liberalisation in tariff policy encouraging imports, and advances in transportation and refrigeration technology to have been complementary factors encouraging the market for meat (1978: 216). For example, when the *ss Strathleven* brought the first really successful cargo from Melbourne to London in 1880, penny ha'penny frozen beef and mutton now fetched fivepence ha'penny at Smithfield (Burnett 1966: 101).

The benefits of the industrial revolution were, however, unevenly spread. The diet of the numerically large working class in the early nineteenth century was as bad as ever it had been and most labourers rarely saw fresh meat at all. A morsel of bacon was luxury, and a farmer might compel his workers to take diseased and unsaleable meat in lieu of wages (Burnett 1966: 14–50, 120). According to Edward Smith's 1863 inquiry into the food of the poor, whilst labourers' diet was barely adequate their family was often malnourished, most food going to the breadwinner (E. Smith 1863). The affluent, by contrast, consumed meat aplenty. *The Family Oracle of Health,* published in 1824, declares: 'It is a bad dinner when there are not at least five varieties – a substantial dish of fish, one of meat, one of game, one of poultry, and above all, a ragoût with truffles' (Burnett 1966: 58). Around the turn of the twentieth century, however, the fashionable meal began to become lighter and more varied:

> the excessive meat-eating of earlier generations was gradually being replaced by dishes of a more vegetarian nature, partly, at least, as a result of the new knowledge of nutrition

Figure 1 Average consumption of meat per head per annum in UK, 1831–1984 (lbs)

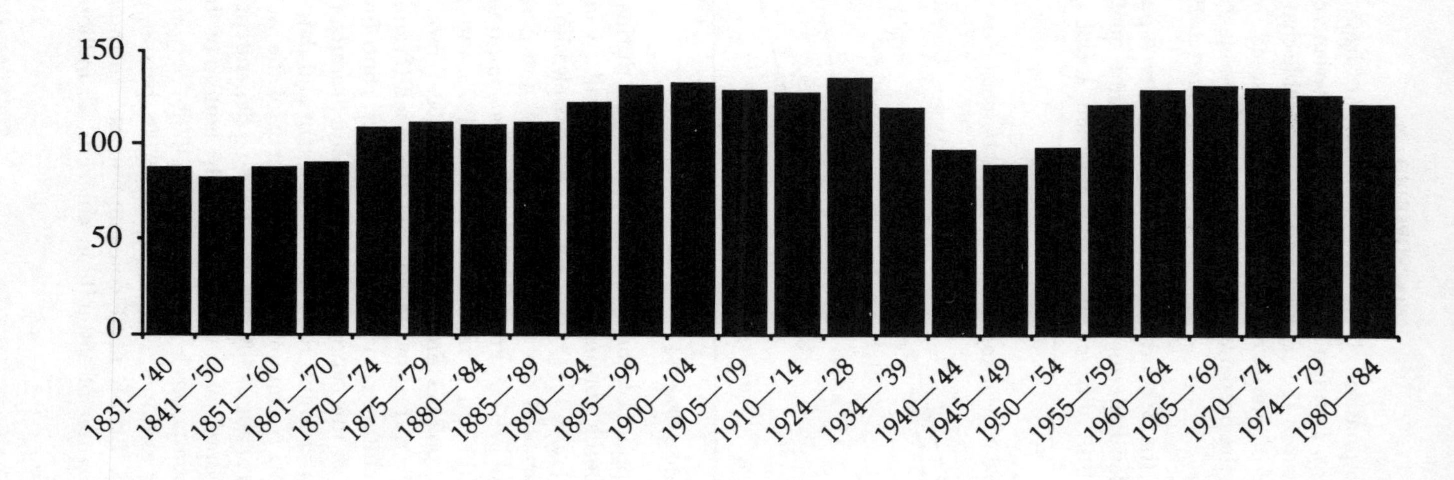

Sources: Perren 1978: 3; Frank 1987

> which emphasised the dietary importance of fresh fruit and vegetables. London had at least two vegetarian restaurants at the turn of the century, and it is noticeable that the later editions of standard cookery books devoted increasing space to the preparation and service of vegetable dishes.
>
> (Burnett 1966: 180)

Prior to the Second World War, average per capita consumption had levelled out but both quantity and quality of meat purchases still differed markedly between social groups (see Figure 2).

Figure 2 Weekly consumption of meat by social class in 1937

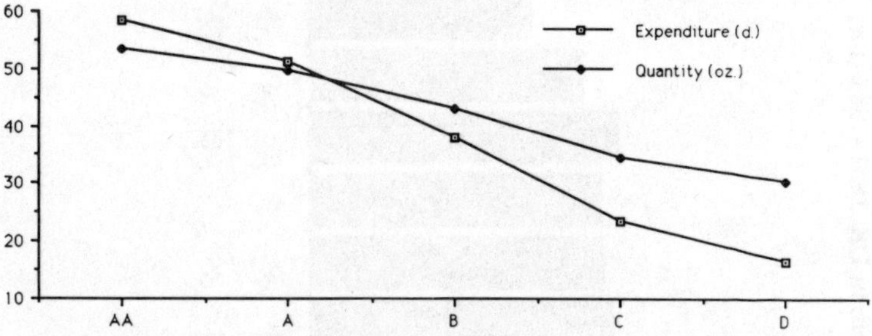

Source: Burnett 1966: 254

Meat supplies reached their lowest level not during the war but whilst rationing continued afterwards, particularly in 1948, 1949 and 1951 (Frank 1987) and demand rose sharply with de-rationing. This rise levelled out again in about 1958 and continued only slowly until 1971. From 1972–1976 consumption began to decline, although it recovered between 1976 and 1979 largely due to EEC policies aimed at reducing the beef 'mountain'. Since then, however, the slow decline has continued (Frank 1987).

UK agriculture remains a major industry, contributing around 2 per cent of the national Gross Domestic Product, of which the meat sector accounts for about 40 per cent and dairying another 30 per cent. It employs about 3 per cent of the workforce, the livelihoods of about 400,000 people being dependent on livestock farming. Nonetheless, the UK lies at the bottom of the European meat consumption league at about 159lb/person/year whilst France, for example, consumes over 220lb/person/year (Sloyan 1985: 3–4). British meat consumption currently demands the annual slaughter of 450 million chickens, 25 million turkeys, 14 million sheep, 13 million pigs, eight million ducks, three million rabbits, and one million quail . . . not to mention quite a few cows. This totals more than two million tonnes of red meat and almost one million tonnes of poultry (Jackman 1989b: 50)

In the two decades, 1966–1986, overall purchases of meat and meat products in Britain fell only marginally, from 37.99 oz/person/week to 37.07 oz/p/w, but within that market occurred significant sub-trends. Beef and veal consumption fell from 8.13 oz/p/w to 6.58 oz/p/w, mostly in the 1980s, and mutton and lamb fell from 6.28 oz/p/w to 3.01 oz/p/w. Pork consumption rose from 2.76 oz/p/w to 3.64 oz/p/w, matched by a fall in uncooked bacon and ham eating from 5.30 oz/p/w to 3.68 oz/p/w. The major rises in consumption were for poultry and cooked chicken, from 4.06 oz/p/w to 7.30 oz/p/w, and for 'other meat products', from 2.78 oz/p/w to 5.67 oz/p/w (Central Statistical Office 1990).

Figure 3 Indices of meat consumption in the home, 1961–1988

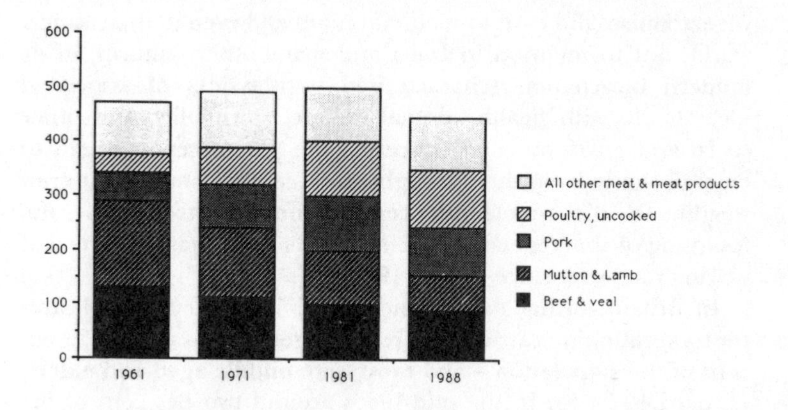

Source: Social Trends 1990: 16

In particular 1980 to 1985 saw beef and veal consumption in the home fall by 20 per cent, mutton and lamb by 27 per cent, and pork by 17 per cent. These falls were balanced by a rise, beginning in the 1960s, in meat eaten away from home, especially in fast food establishments (Sloyan 1985: 4). A significant shift has occurred in meat retailing patterns: between 1979 and 1989 independent butchers' share of the market for meat fell from

56 per cent to 39 per cent while that sold in supermarkets rose from 20 per cent to 38 per cent (*Butcher & Processor* 1989a: 5).

Meanwhile, demand has risen for meat perceived as healthier, such as lower fat products and more white meats, as well as for free-range and organic meat which now constitutes a small, specialised market. About six times as much poultry was eaten in 1984 as in 1954, and in July 1989 the British Chicken Information Service announced that 'Chicken is now Britain's most popular meat – beating beef into second place for the first time' (Frank 1987; *Meat Trades Journal* 1989a: 15).

Regional variations in meat preferences also exist. In the period 1975–1980, for example, Scots consumed about 40 per cent more beef and veal, and 65 per cent more fresh fish, than the British national average, whilst inhabitants of Greater London ate around 43 per cent more mutton and lamb, 28 per cent more pork, 32 per cent more poultry, and 56 per cent more processed fish products (MAFF 1982).

Numbers of vegetarians have also been rising. Although vegetarianism did exist in medieval times and even in the classical world, not to mention in India and many other cultures, in its modern incarnation (characterised by a system of associated ideas to do with health, animal welfare, spirituality, and other social and environmental concerns) the phenomenon began to emerge slowly from the late eighteenth century onwards. It grew steadily in the nineteenth century, notably marked by the founding of the Vegetarian Society in 1847, and has continued at varying rates ever since (Twigg 1983: 20).

In Britain during the Second World War 120,000 applicants for food rationing cards registered as vegetarians – about 0.25 per cent of the population – and most were middle aged and elderly (Erhard 1973: 5). In the mid-1980s around two per cent of the adult population did not eat any meat at all, with a further 2 per cent avoiding red meat (Harrington 1985: 5; Realeat Survey 1986). At time of writing in 1990 this had risen to 3.7 per cent vegetarian and 6.3 per cent avoiding red meat – totalling 10 per cent of the population (5.6 million people). The numbers of vegetarians has risen by 76 per cent since 1984, whilst the number avoiding red meat has more than trebled in that period.

Figure 4 Non-meat eaters. Make-up and change 1984–1990

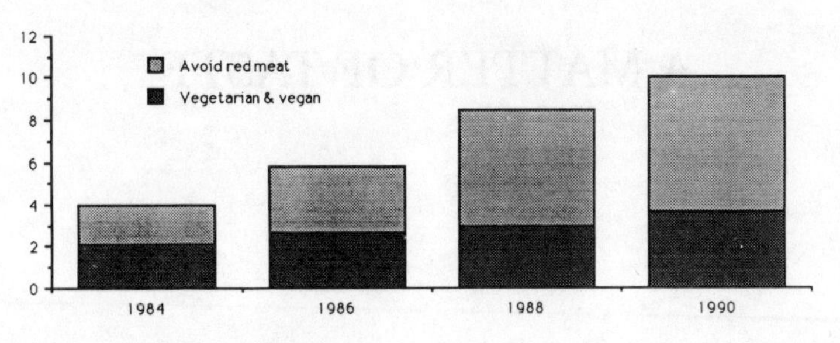

Source: Realeat Surveys 1984–1990

Meat avoidance is more common among women than men – particularly amongst the young. On the most recent evidence 22.4 per cent of 16–24-year-olds and 12.8 per cent of all women eat meat rarely or not at all, compared with 9.1 per cent of young men and 7.1 per cent of all men. Half of all British women claim to be 'eating less meat'. Geographical variations also exist with over 44 per cent of Scots in 1988 claiming to be reducing their meat intake as against 21.5 per cent in Wales and 35.3 per cent in London. In 1990, however, the increase in vegetarianism was greatest in the south of England, where 37 per cent more people were abstaining than in 1988 whereas vegetarianism had apparently decreased by 12 per cent in Scotland. Meat abstention is also more likely to be found amongst the more affluent. Of members of the AB socio-economic groups 4.5 per cent and of C1s 5.5 per cent eat no meat at all, compared with 2.5 per cent of C2s, and 2.9 per cent of DE group members (Realeat Surveys 1988, 1990). This points to one of the most significant aspects of the rapid recent growth in numbers of people avoiding meat: namely that – perhaps for the first time in history – meat avoidance today is often a matter of choice rather than of necessity and is most prevalent among better off and better informed members of the population.

3

A MATTER OF TASTE

The word 'taste' is ambiguous. We use it both to describe the objective flavour of an item as well as to say whether or not we enjoy it – whether, so to speak, its taste is to our taste. These are by no means the same although we do habitually discuss our food preferences as though their origins were somehow inherent in the substance: 'steak tastes good' or 'meat is really satisfying'.

Before taking this discussion of meat further, however, we must dispose of the misleading notion that we crave animal flesh simply for its physical qualities. Just as beauty is said to be in the eye of the beholder so flavour is largely in the tastebud of the consumer – it is a learned reaction. Our attitudes to different foods are conditioned by the associations which we invest in them and we learn these from the day we are born. Mary Douglas notes that:

> Nutritionists know that the palate is trained, that taste and smell are subject to cultural control. Yet for lack of other hypotheses, the notion persists that what makes an item of food acceptable is some quality inherent in the thing itself. Present research into palatability tends to concentrate on individual reactions to individual items. It seeks to screen out cultural effects as so much interference. Whereas . . . the cultural controls on perception are precisely what needs to be analysed.
>
> (Douglas 1978: 59)

As if to illustrate Douglas's case, for example, Paul Fieldhouse states that 'Some foods confer high status on the eaters' (1986: 77) as if high status were somehow a natural component of the food itself. But to cite prestige or good taste to explain food habits is to put the 'cart before the horse' for these are social attributes. George Orwell, similarly, assumes the role of arbiter of taste in noting that we can come to like almost anything:

> [The] English palate, especially the working class palate, now rejects good food almost automatically. The number of people who prefer tinned peas and tinned fish to real peas and real fish must be increasing every year, and plenty of people who could afford to have real milk in their tea much sooner have tinned milk.
>
> (Orwell 1937: 89)

That our likes and dislikes do not greatly depend upon the nature of the foods themselves is clear from the wide variations in people's preferences, both within our own society and between different cultures around the world and in history. One person's meat is indeed another's poison. For example, Patricia Pullar reports that in ancient Rome sow's wombs eaten together with sow's udders were a delicacy (1970: 10n) – a dish at which most modern westerners would shudder. Few of us, however, often ponder the fact that our daily breakfast eggs originate as part of an animal's reproductive system. British supermarkets do not sell dog or horse; nor do we eat many sparrows or larks; nor slugs or grubs; nor marigolds, medlars, or quinces. Some disgust us, others are just not widely viewed as food, although each is consumed elsewhere or was here in the past. Amazonian Kayapo Indians relish termites, and aboriginal Australians witchety grubs; we, generally, do not. Tastes vary and tastes change.

Most adult Britons, including those of us who consider ourselves open-minded, would tend not to eat sweet custard together with beef mince although we might enjoy each individually and though there is little convincing argument against the dish on rational or health grounds. It simply contradicts our normal culinary patterns – our familiar taste. Views differ enormously as to which plants, animals, or other items should be regarded as food, as well as to how they should be presented, as elegantly expressed by a nineteenth-century author:

> Perhaps there is no such thing in persons who are grown up as a perfectly pure and natural taste. The taste may be sound and even fine, but it is always more or less influenced

> by custom and by association, until it breeds an Acquired taste which is not to be reasoned with and will not be denied. The Greenlander takes to tallow; the southern Frenchman glories in garlic; the East Indian is mightily in pepper. No force of reasoning can prove to them that other tastes are better; they have an Acquired taste which insists on being pampered. And precisely the same phenomenon occurs, though in a less marked way, when we get a dish which we know, which we expect, and which does not correspond to its name. A very pleasant Julienne soup can be made without sorrel; but those who look for the sorrel always feel that without it the Julienne is a failure.
>
> (Dallas 1877: 12–13)

We do not even taste things in the same way. In a classic experiment (van Skramlik 1926) volunteers are asked to sample ammonium chloride and then attempt to mix common salt, tartaric acid (sour), and quinine hydrochloride (bitter) to match its characteristic balance of flavours. Results vary widely, some individuals requiring no quinine at all, some ten times the tartaric acid as others. This suggests that, physically, we each respond differently to unfamiliar flavours and that chemical taste does not become perceived taste until we have learned it – or in other words until we have developed an opinion of it.

Taste is not an absolute. It is something we develop whilst growing up within a culture which has its own general preferences. The balance of evidence suggests that humans have an 'early and probably innate preference for sweet tastes', sugar perhaps acting as a sort of biological label for useful calories; negative reactions by the young to sour and bitter flavours may also be of adaptive value since 'most drugs and poisons have a bitter taste' (Cowart 1981: 60). Even these predispositions can, however, be unlearned. Many, for example, grow to dislike sweet foods, just as we can learn to enjoy foods that are fiercely hot, bitter, salt, sweet, or sour – indeed practically any substance that is not excessively toxic. Indeed some foods we enjoy hardly feed us: if we choose to consume diet colas or non-digestible fat substitutes, for example, surely it is not for their nutritional qualities but because, for one reason or another, we have learned to appreciate their flavours.

It is the ideas we have about a food in relation to our entire cosmology – our view of how the world *is* – which govern how it tastes: our view of its edibility or desirability. Custard is part of the sweet course; meat is not. Wine tastes better from a crystal goblet than from a chipped clay mug. As Jean Soler remarks of the semiotics of food in the Bible (1979: 129), the 'explanation of food preferences and aversions must be sought not in the nature of the food items [but in a] people's underlying thought pattern'. An item's edibility depends not upon its flavour but upon its being found a position in our own classification of acceptable foods. When members of the *That's Life* studio audience are fed samples of snail, for example, it is only when its real identity is revealed that they screw up their faces and attempt to spit out the remains (BBC1 14 Jan. 1990). Our reaction is to the image of a food. Thus wholefoods can be just as unpalatable to the habitual consumer of convenience foods as junk foods are to the lover of health foods. Taste is an acquired outlook; it is largely a matter of whether we believe we *ought* to like something:

> 'We've never ever advertised the café as a vegan café, partly because we don't want to put people off – 'a vegan café? I'm not going to like vegan food' – though once they try it I'm sure most people don't even notice.'

The foods we select reflect our thought, including our conception of our actual or desired way of life and our perceptions of the food choices of people with whom we wish to identify. The popularity of cookbooks and branded foodstuffs bearing the name of media personalities with little previous reputation for culinary ability are but one example of this trait. It is not just that our food choices are sometimes influenced by a particular person or group; all of our alimentary behaviour is. We eat nothing in isolation, but as part of our culture – non-conformist habits and changes over time notwithstanding.

Social groupings have characteristic preferences – Barthes (1975), for example, suggests that, in recent years, lower income groups have tended to favour sweet, smooth, and strong flavoured substances, whilst upper income groups prefer bitter, textured, and light substances. The tendency to prefer foods identified with groups to which we belong, or aspire to, and to reject the preferences of reputedly inferior groups, accounts for many fashions in what is popular, prestigious, or pernicious. A much-commented on example is the changing images of white and brown sugar, and of white and brown bread:

> From medieval times, the high prestige of white bread has been well documented in both England and France: the further down the social scale, the darker the bread. The upper classes regarded black and brown breads with aversion – it was even claimed their stomachs could not digest them – while the lower orders aspired to white or whiter bread . . . White bread having become available to all and brown bread having thus, so to speak, fallen off the bottom of the social scale, the brown reappeared towards the top. For the fashion for wholemeal bread has begun to spread downwards from the upper reaches of the social scale since the 1950s.
>
> (Mennell 1985: 303)

We feed not only our appetite but also our desire to belong. Foods express social values, and by consuming them we acknowledge a shared set of meanings. Their rejection can therefore signal dissent – whether by infants, religious sects, or even at the Boston Tea Party.

Familiarity or tradition is commonly said to govern food choice. An American textbook on *Medical Anthropology,* for example, states that 'Tradition is also important in determining diet. Traditional foods become symbols of ethnic identity, and diet can be highly resistant to change' (McElroy and Townsend 1985: 195). Habit also underpins Douglas and Nicod's work on the 'basic English system that underlies regional variations' (1974: 747); every culture, they argue, has a unique meal structure – a frame of rules as to how to construct a 'food event', unconsciously conformed to time and again.

But such accounts ultimately say little more than that we prefer the familiar. This applies not only to particular foods but also to how we order them, how we accompany them, and the variety and variability of our choice. To formalise that structure is of limited use, since habits are by no means fixed. The rules supposedly isolated – of combination of food types, hot and cold, wet and dry, and so on – are themselves as changeable, or unchangeable, as the foods they govern since those rules are cultural:

> 'At one stage I was translating for the Dutch consul, and . . . the Scots wifies, they would get the number from the

> Consul, phone me, and say give me a Dutch menu. And then I would say – such and such a soup, and they'd say "Oh, that's lovely! And I'll put a little bit of this or that in" . . . and I'd let that one pass. And then you gave the main course, and the question would be "Would it matter if it was mashed potatoes – and beef instead of pork?" "Yes it would! Because the whole idea is that . . ." And by the end they had such lovely British menus! And they'd hang up the phone and I'd think, what the hell am I doing?'

The range of ingredients and repertoire of recipes familiar to a particular cook is likely to limit change to an extent since the preparation of well-known foods is clearly easier than constant experimentation, and since having safely consumed something once we can be reasonably confident that it is not intrinsically harmful. Experiments have demonstrated, for example, that although capable of adapting to new foods rats will normally adopt them only if familiar foods are unavailable and then only slowly. Whilst it would be wrong to suggest that our behaviour is necessarily the same as that of the rat it is also perhaps arrogant to hold that we are entirely free of biological influences. Of humans, too, Simoons notes that:

> The suggestion that unfamiliarity with an animal may contribute to the rejection of its flesh appears to have considerable merit. An 'emancipated' Westerner will often refuse the flesh of strange animals, partly perhaps from fear that it will cause illness but also from reluctance to partake of something new.
>
> (Simoons 1967: 112)

Although apparently rational, Simoons's circular explanations tell us little. Why should the westerner suspect an unfamiliar animal to be less healthy than a familiar one if it is known that others enjoy that animal's flesh? Reluctance to partake of the new is self-evident, and little more. Simoons is unwilling to recognise our own behaviour also to be influenced by rejection of the unknown, even if we represent that rejection to ourselves in logical terms such as of contagion. In discussing less 'emancipated' cultures, Simoons ventures further:

> Primitive man views the flesh of unfamiliar animals with even greater trepidation, for it may be the means by which

> harmful spirits or other mysterious elements enter his body . . . The fear of eating an unfamiliar animal is frequently increased through its introduction by a disliked and feared donor group, who might do serious harm or might be regarded as unclean . . . [For example] certain Andaman Islanders . . . will not eat particular foods when they are away from their own sections of the islands, perhaps from fear that in a strange place the chances of illness are greater and the spirits are more dangerous.
>
> (Simoons 1967: 112)

Unfamiliar foods fail to offer the same security since we are as unsure of their symbolic status as we are of their physical safety. The danger, in other words, is to our minds as well as to our bodies. In this respect British Islanders are remarkably alike the so-called primitive Andaman Islanders discussed by Simoons, tending to adhere to the culinary patterns of the culture with which we identify:

> [Of self-catering British holidaymakers visiting Spain in 1987] nearly two thirds did not buy any food that they would not normally eat at home, and just under half preferred to eat in their villa to local restaurants . . . Nearly everyone shunned locally-bought bacon and eggs, or cold meats, for breakfast: only one in 20 overcame their nerves and gave them a try.
>
> (Reeves 1988: 1)

This provides a more helpful clue to the danger associated with strange foodstuffs. The threat manifestly comes not only from direct contagion, but also from association with unwelcome ideas – in this case particularly with strangers. Habit and traditions offer affinity with our peers and continuity with our individual and collective past. The benefit of familiarity or convenience is ideological as well as practical: we need not repeatedly tackle questions of classification, or confront ideas which might disturb and deter us.

Thus, if a tradition persists, it is probably because the ideas embodied are still valued. If the taste for a particular food, such as meat, remains strong, then efforts will continue to be devoted to securing its regular supply. If less effort is made, it may indicate that different ideas have come to prominence. Traditions, whether those of the individual, the family, or the nation, can and do mutate, just as habits can be broken. According to Markey (1986), for example, when English settlers colonised the Swan River in Western Australia in 1829 their diet, without the complication of contact with other cultures, was remarkably resistant to change in spite of the enormous change of their environment. It remained a stable element of culture until after the Second World War. Since the 1960s, however, dramatic changes have taken place in Western Australians' food habits, at a time, Markey suggests, when the society was changing its mind about other matters too. Food, thus, may be not only a 'cultural spoor' but also a cultural omen: a straw that shows which way the winds of change are blowing. George Orwell indeed thought 'it could be plausibly argued that changes of diet are more important than changes of dynasty or even of religion' (1937: 82).

The revived popularity of wholemeal bread mentioned earlier, for example, needs more explanation than the cyclical dynamics of competing social classes seeking to outdo or to emulate each other. Rather, it should be seen against the background of a general movement away from highly refined products towards more wholesome foods perceived as healthier – including not only brown bread, but also, for example, brown rice, and even brown sugar. This can be interpreted as an intimation of new intolerance towards the excessive industrialisation of food supplies, and perhaps of modern life in general. 'Naturalness' is part of the signification of wholefoods in general, and wholefoods have seen a distinct surge in popularity in recent years. Unadulterated purity is a value that has been coming back into fashion, and society has been learning, once again, to appreciate its flavours.

4

MORE THAN A MEAL

All over the world food means much more than mere nutrition. Perhaps it is singled out for such significance because everybody, everywhere needs to eat. Perhaps it is because – along with only a few other similarly significant acts such as sex and defecation – eating breaches our normally sacrosanct bodily boundaries. Maybe it is important that when we consume we literally incorporate into our own bodies the physical material – and possibly the spiritual essence – of other animals and of the outside world in general. But whatever the reason, we routinely use food to express relationships: amongst ourselves and with our environment. The obtaining and sharing of food can be an eloquent statement of shared ideology and as such expresses group affiliation and apparent solidarity. W. Robertson Smith noted that 'those who eat and drink together are by this very act tied to one another by a bond of friendship and mutual obligation' (1889: 247); Radcliffe-Brown held that for the Andaman Islanders 'by far the most important social activity is the getting of food' (1922: 227); and Darlington (1969) suggests that commensality may be the most important basis of human associations. Food is a system of communication, a body of images, a protocol of usages, situations, and behaviour (Barthes 1975):

> Food is prestige, status and wealth . . . It is a means of communication and interpersonal relations, such as an 'apple for the teacher', or an expression of hospitality, friendship, affection, neighbourliness, comfort and sympathy in time of sadness or danger. It symbolises strength, athleticism, health and success. It is a means of pleasure and self-gratification and a relief from stress. It is feasts, cere-

mony, rituals, special days and nostalgia for home, family and the 'good old days'. It is an expression of individuality and sophistication, a means of self-expression and a way of revolt. Most of all it is tradition, custom and security.

Different foods satisfy these needs and beliefs of people in different cultures. Some foods are linked to the age and sex of the individual . . . There are Sunday foods and weekday foods, family foods and guest foods; foods with magical properties, and health and disease foods.

(Todhunter 1973: 301)

'It sort of came as a final rejection of her cooking. It was actually at that level, when I was at home at the weekends. She was cooking lovely meals, and I was not eating the meat, which everyone else was enthusing over: it was "a beautiful piece of roast you've got this weekend", and "oh, it's lovely tender lamb for this time of year". And I was saying that I don't want it. And it was – it was like rejecting part of her . . .'

Meat is just a way of life for British families

Sir – You may not consider it very important, but I'd like to tell you a simple story about a piece of pork.

Myself and my family bought it on Saturday for our Sunday lunch.

It was an attractive joint and, despite some good Saturday night TV, the prospect of the meat remained in our minds and there was a hint of expectation on all our faces.

For three hours on Sunday the smell of it cooking practically drove us wild. When we finally sat down at the table, all the troubles of the week seemed to drift away at the prospect of a delicious family lunch.

The meat was wonderful and I thought how much the £5 joint had contributed to this typically British, family scene. My family left the table feeling well-fed and happy and the cold meat made a meal on Monday night as well.

Meat is not just a meal, it is a way of life.

T. Cook. Basildon, Essex.

(*Meat Trades Journal* 1987: 2)

Conversely, those who diverge from community standards are commonly stigmatised since their dietary non-conformity is (correctly) taken to indicate broader differences in values. Consider, for example, the dismissive tone and marginalising vocabulary in an American sociologist's treatment of alternative diets:

> In studies of social movement and the formation of sects and dissident groups, the role of food cannot be underestimated. In adhering to some dietary rules, what to eat, when to eat, or when not to eat, groups maintain control over their members. They also require members to deviate from the general population when they venture outside their group. This behavior is one of the most effective ways of assuring adherence to special group codes. Vegetarianism, which has recently attracted a variety of individuals and been intimately connected with several modern movements (Barkas 1975) might serve as an example. It is hard to find a common denominator among these groups, except that they are all in some way intent on establishing a difference and attracting attention to it. A more recent phenomenon seems to be the fad for 'natural foods,' which often corresponds to some political and social dogma and thereby serves to bring adherents together.
>
> (Back 1977: 31–32)

Back seems desperate neatly to categorise all such individuals according to their culinary deviation from the mainstream. There are overtones of conspiracy theory with groups 'maintaining control' over members who must conform to 'some political or social dogma'. The disparaging use of terms such as fad and allusion to brainwashing in such value-loaded description implies the majority diet to be rational. But Back's dogmatic zeal obscures from him the diversity of ideas and meanings involved, not only in the minority groups he isolates (many of whom may indeed deliberately be rejecting what Pope called 'carnivoracity') but also within the dominant culture with which he evidently identifies. He assumes that vegetarianism ought to mean much the same thing to all vegetarians and is perplexed since the only common attribute he can advance is their 'difference' to which he weakly and unjustifiably ascribes a ubiquitous desire to attract attention. The seemingly bizarre food habits of exotic cultures or of minority groups may encourage us to look for extra meanings – of something communicative. But our own are just as expressive, and may seem equally strange to outsiders.

Foods do not intrinsically symbolise. They are used to symbolise. For example, when we are told that the cooked dinner of meat and two vegetables symbolises the woman's obligation as homemaker and her husband's as breadwinner in South Wales (Murcott 1982), clearly the food does not itself stand for home-making and caring. The values are their respective gender roles, whilst the food is the medium through which that is communicated. This function of food is evident throughout our society.

So why is meat a 'Natural Symbol'? What does 'symbolic' in fact mean? The term is commonly used to imply communication that is distinct from direct, literal, discourse. Objects, ideas, or actions, are called symbolic when they represent something beyond their obvious identities. A flag is symbolic because it is more than a piece of decorative cloth; a dove because, by common consent, it represents the concept of peace. As Anthony Cohen puts it, 'When guests compliment their host for having laid a good fire, they probably mean more than that he has achieved a commendable thermal output for a given quantity of fuel' (1986: 3).

Meat certainly qualifies as symbolic in these terms since its economic and social importance is frequently greater than might be anticipated from its purely nutritional value. But symbolism is more than a few ethereal associations which somehow disrupt our otherwise rational judgement. That is only the surface of an infinite system of thought that can be implicit or explicit, private or public, tacit, or overt. Nutrition and economics, for example, are themselves symbolic (Sahlins 1976). As Douglas and Isherwood have argued, 'the problem goes so deep that nothing less profound than a corrected version of economic rationality is needed' (1980: 4). In fact all our goods are of communicative, at least as much as of utilitarian, value – they are good(s) to appreciate (and to be seen with by others who share our view of their value) as well as to consume.

When nutritionists or policy-makers discuss the energy, fat, or protein contents of foods, for example, and expect a willing public dutifully to adapt their habits, they are deceiving themselves in failing to accommodate the numerous other roles that foods play in people's lives. Indeed some 'authorities' seem to live in a fantastic science-fiction world, alien to those of us who simply enjoy buying, preparing, and eating food:

> Another transatlantic viewpoint was given by Dr Carl Unger, a meat consultant based in Georgia, USA, who projected a picture of computerised shopping in the mid-1990s.
>
> He predicted that the shopper would use a computer to work out daily dietary and nutritional needs. These would be translated into a number of dishes and meal components which would be ordered from the local supermarket.
>
> (*British Meat* 1987a: 3)

We routinely stress meat's scientifically recognised function in terms of health and nutrition as the principal determinant of its status. From this, it is generally assumed, its value as an item of economic exchange is derived. Occasionally a certain symbolic importance may also be recognised, such as the macho steak which, like the macho car, may be purported to denote sexual prowess, or such as the view that its importance or prestige could be a sort of social relic from when we hunted to survive or from royal hunting in Days Of Olde.

But there is more to meat than protein. The Shorter Oxford English Dictionary, for example, defines meat as 'The flesh of animals used for food' and a few subsidiary and archaic variations; not even its common usage as a synonym for 'essence' or 'substance' is mentioned (e.g. 'The meat of the argument'). Whilst meat as food may be its principal accepted meaning, this falls far short of conveying the elusive depth of its signification. To one person the word *steak* might suggest a substance whereby affluence or culinary skill can be demonstrated; it might conjure up memories of celebratory dinners by candlelight; it might reassure about the consumption of good, body-building nutrition. To another, steak might stand for cruelty and nausea, with thoughts of horrific conditions in which animals are bred and slaughtered, and images of violence, blood, brutality. No two people will find quite the same meanings in the same word, and no such association can be called 'incorrect'.

The associations can be seen as relative, or 'symbolic truths', since it is upon belief and image that individuals base their actions, not absolute fact. This can help explain apparent contradictions in our behaviour. For example, amid increasing distrust of factory food production and of fatty meats, pork and chicken – in spite of being intensively farmed and of sometimes being no lower in fat than many 'red' meats – sell relatively well, partly because they are less bloody and have been marketed as white meat, and therefore less graphically meaty. Game and salmon are similarly favoured by many as 'wild' even though a high proportion of purchases are in fact of farmed stock.

Each meaning, and countless others, is true for the individuals concerned, extending the significance of the name of a particular meat, or of meat in general, far beyond its function as a foodstuff. It is the totality of these ideas which combine to form a language, and which constitute culture. We must look to the depth and diversity of meaning in all our ideas and actions. Thus economics and technology are symbolic in the very same sense as unicorns or religious icons, though they undoubtedly differ in both their form and content. And meat carries many meanings.

Sperber (1975) argues that symbolism deals with things which cannot be dealt with rationally (and so incorporated into everyday speech). Sometimes, however, symbols rather seem to concern things which simply *have not been* so classified and expressed, even though they might be. That a particular meaning associated with a symbol is not normally articulated may be because it is all the more persuasive for being unvoiced.

This is perhaps best illustrated by example. Advertisers and marketing executives are well aware that people buy more than utilitarian objects. Successful sales people offer 'lifestyles' and emphasise aspirations and associations rather than function. Much modern advertising, indeed, scarcely mentions the product. For example (excluding the purely image-building genre), a motor car is typically marketed in terms of comfort, economy, power ('to cope with emergencies'), and prestige, often with the implied promise of sexual success for the probable male purchaser, but these messages will be manifest in different forms. The most explicit messages will tend to concern the reputedly rational considerations that the potential buyer will happily acknowledge. Other more emotive meanings, perhaps concerning the social or business niche to which the target is characterised as aspiring, are unlikely to be spelled out in the text but are nonetheless clearly written in the situations portrayed.

Barthes (1975) identifies three themes that stand out in food advertising – the commemorative, eroticism, and health – arguing that this reflects the collective psychology more than it shapes it. But whether to sell the chicken or the egg, it is clear that much of advertising's symbolic power lies in its unvoiced suggestion. The message is altogether more compelling as part of the taken-for-granted context rather than as an overtly propagated claim. Indeed, an idea can have very different meanings depending upon whether it principally circulates explicitly or implicitly, perhaps because only when brought to awareness does it become open to criticism. The motorist who tacitly enjoys the endemic association of sports cars with sexual virility, for example, might feel considerably less comfortable when openly mocked by others for his 'penis substitute'. And were the *reasons* for the association between car and sex, to be more precisely rendered, they would be open to discussion, demystification, and dispute, and might rapidly lose their effect. The power of the ideas depends upon their being communicated without being rendered explicit, for their meaning can then be understood by all concerned, but at the level of assumption, common sense, and accepted fact.

That animals are killed for humans to eat meat is obvious to the point of banality. However, the inherent conquest is rarely discussed overtly in the context of food provision. Our willingness to eschew confronting certain aspects of meat's identity is more than a matter of preferring to sidestep that which might be unsavoury. The fact that most of us make little mention of the domination inherent in rearing animals for slaughter does not indicate that it is irrelevant. On the contrary, that which remains unsaid about meat conveys an added dimension of meaning which is particularly potent. It is the very taken-for-grantedness of values implicit in the meat system which makes the message so powerful, whilst rationalisations of meat's importance (such as nutrition and prestige) partly serve to obscure these values from our consciousness.

Paradoxically, this obscurity preserves and perpetuates the influence of these implicit meanings since, not being recognised, they can scarcely be challenged. Veal, for example, enjoyed high prestige for many years partly, I suggest, *because of* the extreme subjugation of the creatures intrinsic to its production. That, however, was seldom voiced; instead its value was explicitly attributed to such arbitrary qualities as delicate flavour and light colour. Only once its production methods were brought into the domain of explicit public consideration did they become intolerable, at which time this previously inherent meaning lost its positive power and instead became a negative influence on the meat's popularity.

What was true for veal yesterday could also become true for meat in general tomorrow. In the chapters that follow I shall show that the unvoiced symbolic values which continue to underpin meat's popularity today principally concern our relationship with nature, as we perceive it. In this way changing attitudes to meat, as revealed by changing habits, may also be eloquent commentary on fundamental developments in society. Meat's signification, I suggest, principally relates to environmental control, and it has long held an unrivalled status amongst major foods on account of this meaning. But meat's stature is not inherent in its substance, but has been invested in it by successive generations who highly valued its meaning: who *liked* the notion of power over nature that it embodies. Its waning prestige – and outright rejection by many – may be indicative of more than changing tastes in food. Thus meat is more than just a meal; it also represents a way of life.

Part II

MEAT IS MUSCLE

5

EVOLUTION AND ELEVATION

Human groups the world over, according to Maurice Bloch, are concerned to resolve deep-rooted questions of identity – questions which commonly focus upon one particular problem:

> how far is man separate or continuous with animals, plants and even geographical and cosmological events? The answer is, like any answer to this fundamental question, always unsatisfactory, and therefore such answers endlessly throw up further problems, thereby initiating an ongoing, never resolved, discourse.
>
> (Bloch 1985: 698)

Analysis in the social sciences commonly centres on a putative distinction between nature and culture: between that which would (presumably) happen if we behaved as just another ape, and that which we do create from nature's potential, whether consciously or otherwise. Some believe such a distinction to exist in all human thought. Sherry Ortner, for example, contends that 'We may . . . broadly equate culture with the notion of human consciousness (i.e. systems of thought and technology), by means of which humanity attempts to assert control over nature' (1982: 490). A favourite illustration for Lévi-Strauss is the activity of cooking which, though not strictly necessary, is apparently pan-human, suggesting that it may be a universal symbol 'by which culture is distinguished from nature in order that men might reassure themselves that they are not beasts' (Leach 1970: 129).

Strathern, on the other hand, has complained that 'there is no such thing as nature and culture' (Strathern 1980: 177), basing her argument on the inappropriate use of these terms, since the Hagen of Papua New Guinea (for example) clearly do not see 'nature' and 'culture' as we do. More damningly, Stephen Horigan (1988) has shown the nature–culture opposition to be something of an intellectual straightjacket, shackling discussion to sterile conceptions of human uniqueness and the autonomy of culture. Unless we are willing to recognise other species also to have learned cultures, and ourselves to be animals – not gods or fairies (Midgley 1980: 55) – we easily lose sight of the need to pose new, interesting, questions:

> We should not be trying to 'prove' that chimpanzees have or do not have human language, but rather to understand what the ape's undoubted ability can tell us about the nature of language and what further insights it can give us about the mind of the ape.
>
> (Horigan 1988: 106)

Be that as it may, western thought has experienced a magnetic attraction to the quest for an attribute that uniquely distinguishes us from the rest of creation, and for an epoch when recognisably human beings first emerged. In addition to Lévi-Strauss's championship of cooking (like Boswell before him), Aristotle described humans as political animals; Thomas Willis called humans laughing animals; Benjamin Franklin, tool-making animals; and Edmund Burke, religious animals. 'What all such definitions have in common is that they assume a polarity between the categories "man" and "animal" and that they invariably regard the animal as the inferior' (Thomas 1983: 31).

Nature is conventionally regarded as not only other than culture, but contrary to culture. Human society is not only distinct, but distinctly superior. The material world has traditionally been literally beneath our dignity – a resource for us to use at our will, but towards which we have little obligation, if any, to reciprocate in return for whatever we extract. It has, indeed, been almost a divine duty to subdue the wild. The influence of this intellectual heritage persists to this day throughout our media of discourse, and is well exemplified by how we utilise the world's natural resources . . . not least the flesh of other animals. The 'nature–culture dichotomy is so prevalent in contemporary Western thought that we tend to take it for granted without seriously beginning to challenge it' (Birke 1986: 102).

Bullfighting provides a spectacular example of the ostensibly primordial struggle between culture and nature, formalised in ceremonial battle. According to Garry Marvin this Spanish institution survives as a popular arena in which the superiority of human abilities and intelligence over the might of brute nature is ritually certified.

> The cultural significance of the corrida [bullfight] can be best understood if it is interpreted as an event which both encapsulates and succinctly and dramatically summarizes the important structural oppositions of nature and culture which underlie the idea of what it means to be civilized or truly human.
>
> (Marvin 1988: 128)

The entire pageant is constructed to articulate the inevitable victory of brain over brawn, qualities that are regarded as singularly human and animal respectively. By definition, a good bull must lack human intelligence:

> A bull which waits before it charges, which looks from the cape to the body of the man and then back again, is not a bull with *nobleza* but rather it is one with *sentido* (sense or judgement); it appears to be weighing up how to attack. A bull with *sentido* is difficult to perform with, because the *torero* cannot be certain what the animal is going to do. The whole basis of *toreo,* which is that the man deceives the bull, breaks down if the bull attempts to deceive the man as well. A bull with *sentido* is not a good bull because it implies decision and judgement on the part of the animal; qualities which should only apply to men.
>
> (Marvin 1988: 103)

The bullfight is just one of the myriad ways which human societies have devised to bolster our perceived superiority over lesser beasts, even as we censor inconveniently contrary evidence. In the western world, we have consistently represented our environment as a threat to be conquered, a wilderness to be tamed, a resource to be utilised, an object with few intrinsic needs or rights. This ethical position has come to be something of an ideological imperative, endemic in religious, academic, scientific, commercial, popular, and mythological forms.

We still rear our children on countless fairy tales in which the deep, dark woods are a symbol of untamed danger – the lair of wicked witches or savage and cunning wolves waiting to pounce upon the innocent. Or we relate stirring tales of brave pioneers venturing into the harsh wilderness to extend the boundaries of civilisation, felling forests to construct log cabins as safe havens from the outrageous dangers of the wild, planting crops in the untilled ground and gradually bringing human organisation where once there was natural chaos. (Science fiction, such as the popular American television series *Star Trek,* infinitely extends our horizons to create a legendary future where our mission is 'to boldly go' to extend a distinctly human order ever wider into the furthest reaches of space – 'the final frontier'.) According to historian Roderick Nash, this was indeed how the average American colonist viewed nature:

> Wilderness . . . acquired significance as a dark and sinister symbol. [The pioneers] shared the long Western tradition of imagining wild country as a moral vacuum, a cursed and chaotic wasteland. As a consequence, frontiersmen acutely sensed that they battled wild country not only for personal survival but in the name of nation, race and God. Civilizing the New World meant enlightening darkness, ordering chaos, and changing evil into good.
>
> (Nash 1967: 24)

The roots of such attitudes run deep. For many years religion and philosophy provided our answers to this most compelling question. Arthur Lovejoy suggests that the concept of the Great Chain of Being – a notional hierarchy of all creation ranging from God at the apex, through the angels, to humans, to other animals, to plants, and to the inanimate – has been:

> one of the half-dozen most potent and persistent presuppositions in Western thought. It was, if fact, until not much more than a century ago probably the most widely familiar conception of the general scheme of things, of the constitutive pattern of the universe; and as such it necessarily predetermined current ideas on many other matters.
>
> (Lovejoy 1936: vii)

The central principle, which underpins, for example, commonplace references to higher and lower animals, can be traced back to Aristotle who suggested that:

> we may infer that, after the birth of animals, plants exist for their sake, and that the other animals exist for the sake of man, the tame for use and food, the wild, if not all, at least the greater part of them, for food, and for the provision of clothing and various instruments. Now if nature makes nothing incomplete, and nothing in vain, the inference must be that she has made all animals for the sake of man.
>
> (Aristotle 1984: 1933–1934)

Lovejoy charts the progress of the idea through western history and thought, including the philosophy of Aquinas, Leibniz, and Spinoza, the science of Copernicus, Kepler, and Bacon, and the art of Milton, Pope, and Victor Hugo. Paradoxically, however, it was in the eighteenth century, when speculative metaphysics was already waning in favour of triumphant empiricism, that the conception of the universe as a Great Chain of Being reached its widest diffusion and acceptance (Lovejoy 1936: 183). The explanation for this incongruity is straightforward. It is that, while the scheme notionally testified to the existence and glory of God, its clear effect was the glorification of humanity, since our median position was supposed to mark the transition from mere sentience to intellectual forms of being. This philosophical scheme, though doubtless believed by its adherents to have been arrived at by rigorous metaphysical enquiry, reflected and served the wishes of the culture which gave it birth and nurtured it to maturity:

> Today when our ascendancy over nature seems nearly complete, there are plenty of commentators ready to look back with nostalgia at earlier periods when a more even balance obtained. But in the Tudor and Stuart age the characteristic attitude was one of exaltation in hard-won human dominance. Man's dominion over nature was the self-consciously proclaimed ideal of early modern scientists.
>
> (Thomas 1983: 28–29)

The concept of the Great Chain of Being supported the assumption that all material things exist for the sake of humans: a proposition vital to the scientific and industrial endeavour. Certainly the expansion of economic activity in the sixteenth century ran parallel to the increasing influence of mechanistic philosophies, perhaps partly due to the potentially restraining effect that older 'organic' philosophies could have had on exploitation of the world's physical, animal, and human resources (Merchant 1982). Like a self-fulfilling prophecy, the ethic of human supremacy was consistently invoked to justify the steadily increasing power of industrial civilisation. Conversely, the exploitation of nature tangibly exemplified, and so was seen to confirm, our evident difference and natural elevation.

It is thus appropriate that meat consumption rose rapidly in the early modern period, when the industrial revolution was at its height, and when such views were most clearly and unambiguously expounded, vindicating untrammelled application of the scientific method and of technology, and extolling human triumph. Meat was, for the more powerful members of society at least, to be provided in lavish quantities for enjoying to the full. Thomas (1983) reports that the seventeenth-century philosopher Henry More, for a time 'the most zealous defender' of the notion of the Great Chain of Being (Lovejoy 1936: 125), was sure that cattle and sheep had only been given life in the first place to keep their meat fresh 'till we shall have need to eat them' (More 1655: 116); William King, writing in 1731 on the origin of evil similarly argued that there is no injustice in killing oxen as food for a 'more noble animal' since it is only for that that they are given life (King 1731: 118–119).

Notions of the dualism of mind and body and of the mechanistic operation of living things, distilled most famously by Descartes but with a considerable intellectual heritage, were eagerly adopted to sanction already rapidly developing technological investigation and exploitation of our living and non-living resources. Meanwhile, Locke's empiricism supported belief in the possibility and value of objectivity – the attitude of regarding everything external to the self as an object – and the need for dispassionate observation (Novak 1971). Science and technology blossomed in the spirit of detached curiosity, if not outright hostility, with which we became accustomed to view our surroundings, their advances bringing what we learned to regard as civilised benefits, at least for the fortunate. Material progress had the additional effect of inducing others to adopt similar ideals so that western scientific, industrial, and commercial values today inflect the thinking and influence the activities of most of the world's population. On many fronts, the feeling of involvement and intimacy with the natural environment was lessening, or, in other words, the conceptual divide between nature and culture was widening.

Homo sapiens sapiens eats more meat than any other primate. Indeed, until quite recently it was widely believed that we were the only primate to eat meat at all. This, however, is untrue. Other apes do eat some meat, though considerably less frequently and in lesser overall amounts than do most humans. It provides them with an occasional source of rare nutrients (Goodall 1965; Teleki and Harding 1981).

That we so readily presumed non-human primates to be vegetarian is itself significant, for meat eating is one key trait by which, traditionally, we have characterised ourselves as human. In the history of our post-biblical attempts to isolate the emergence of our species, hunting is perhaps the single most widely cited attribute. A modern writer on deer hunting, for example, casually defines the birth of humanity by the beginnings of skilled hunting, and progress by the development of more efficient hunting technology:

> The basic principle, from the time that Neolithic Man, feeling peckish no doubt, first stalked a Mastodon . . . has been to get as close as possible to one's quarry to be able to kill it with the minimum of risk to one's self. Man, ever a hunter for food, used his mental skills to help him and developed his weapons accordingly . . . Of course, once gunpowder came on the scene, the picture changed. Prior to that taking deer was by driving to traps or to an enclosed place where they could be slaughtered (horribly reminiscent of the Faroes whale killings), or by hunting with hounds. But now, Man could resuscitate his stalking skills.
>
> (Bowser 1986: 22)

This short reference to when Man 'first stalked' game suggests: that until an indeterminate point in prehistory our forebears were largely vegetarian; that hunting was precipitated by shortage of food; that stalking skills were an important advance in the evolution of the species to a higher level of civilisation; that hunting is properly a male pursuit; and that modern hunting essentially perpetuates timeless tradition. As usual, and notably, the writer does not elaborate upon these 'facts' which are, after all, common knowledge.

Archaeological evidence is interpreted to suggest that pre-*Homo sapiens* groups moved from primarily foraging towards hunting between two and four million years ago (Lancaster 1968) and that many of the major social, biological, and technical developments by which we characterise modern races, such as tool-making and the sexual division of labour, also began around then. The bare facts of critical changes in human activities and capabilities during one period of prehistory are not in dispute. What is interesting, however, is why we believe these developments took place then, and the implications we perceive for the relationship between humans and their environment:

> Hunting changed man's relations to other animals and his view of what is natural. The human notion that it is normal for animals to flee, the whole concept of animals being wild, is the result of man's habit of hunting. In game reserves many different kinds of animals soon learn not to fear man, and they no longer flee . . . Prior to hunting, the relations of our ancestors to other animals must have been very much like those of the other noncarnivores. They would have moved close among the other species, fed beside them, and shared the same waterholes. But with the origin of human hunting, the peaceful relationship was destroyed, and for at least half a million years man has been the enemy of even the largest animals. In this way the whole human view of what is normal and natural in the relation of man to animals is a product of hunting, and the world of flight and fear is the result of the efficiency of the hunters.
>
> (Washburn and Lancaster 1968: 299)

The usual claim is that hunting stimulated and necessitated radical changes in human abilities and organisation, as if social conditions, and even thoughts and ideas, can be straightforwardly read from the meagre flotsam and jetsam of long-extinct communities. We date the genesis of civilisation to dominion over the wild, non-human world. Hunting, the argument goes, may have been even more significant in the evolution of modern humanity than the later development of agriculture. 'Agricultural ways of life have dominated less than 1 per cent of human history, and there is no evidence of major biological changes during that period of time', which is why 'the consideration of hunting is so important for the understanding of human evolution' (Washburn and Lancaster 1968: 293, 294). The popular *Hamburger Book* also tells its audience that progression to meat-eating characterised emergent humanity:

> Meat has been a favorite food of mankind for something like 4 million years. Although archaeological findings have revealed that pre-man (a species of advanced ape) existed as early as 14 million years ago, those creatures are believed to have been vegetarian . . . Then, in about 4 million B.C., a transition phase began during which the advanced ape began to develop into composite ape-man . . . learning to kill for his food with rocks, stout tree limbs, or whatever other natural object lay about him. He was on his way to becoming a habitual meat-eater.
>
> (Perl 1974: 13)

Robin Fox promotes a similar vision of the early origins of humankind, attributing the unequal relationship between the genders to a quest for large amounts of meat:

> the change from sporadic meat eating to a diet incorporating more than 50 per cent meat meant a radical change in the relations between the sexes and between the old and young males. It is these changes that *created* man as we know him . . . In the 'winner-take-all' type of competition sheer strength is what counts; in the primate 'hierarchical' competition it is more control and timing; in the hunting situation it is obviously the ability to provide meat – to provision the females and children. But it is much more complex than this: strength, control and hunting ability cumulate in importance, but many other qualities must accrue to a successful dominant male in a co-operative hunting society. Leadership, organizational ability, and even such burgeoning talents as eloquence, shamanistic skills, etc. eventually come to characterize 'dominance' and hence breeding advantage . . .
>
> (Fox 1985: 8–9)

Serge Moscovici too, in *Society Against Nature* (1976), argues that the seminal change from a gathering to a hunting economy was also the origin of gender inequalities:

> Man made himself into man when he set himself up as a hunter; in other words, when he tried to acquire definite skills and means in order to relate to a given environment and was thus genetically, socially and technically transformed . . . Such a definition is all-encompassing. It can be

> historically situated; it by-passes the restrictive preoccupation with tools and sustenance; and it links up with myth, ritual and individual emotional and intellectual interactions. It also includes men – predacious or hunting – of various species which once coexisted, such as *Australopithecus robustus* and *Homo habilis;* or succeeded each other, such as *Homo erectus* and *Homo sapiens.* It distanced and distinguished them from primates, not because the latter were primates, but because they were wholly dependent on vegetation.
>
> (Moscovici 1976: 30)

The orthodox assumption in each passage is that hunting inaugurated changes in human relationships as well as between humans and their environment. The process is perhaps best described in Washburn and Lancaster's influential article *The Evolution of Hunting* in which they advocate co-operation in hunting, butchery, and war as the source of male–male associations (1968: 295–296). Such viewpoints typically involve an assumption of causality: namely that hunting not only *marked* the appearance of human society, complete with gender differences, but actually *fashioned* it.

And just as hunting can be said to be the fountainhead of human biology, society, and technology, so the advent of farming may be held to be crucial in the development of civilisation, and control of nature is again at the heart of its definition. 'The origin of agriculture' we are told 'was not an instant, chance discovery. Over tens of thousands of years people observed their environments, performed experiments, and gained skills and knowledge to produce a more stable food supply by controlling plants and animals' (Harlan 1976: 89). 'Domestication of food animals occurred quite early in the Agricultural Revolution . . . The purpose of domestication was to secure animal protein reserves and to have animals serve as living food conserves' (Dando 1980: 23).

Sherratt adds that the development of improved animal breeds – an aspect of what he calls the 'secondary products revolution' – was a significant innovation in the development of complex economies, contributing to the process of intensification of production as one of the necessary technological advances (Sherratt 1981). This same theme was pursued by Leslie White, who argued that the neolithic period saw a 'great advance in cultural development . . . as a consequence of the great increase in the amount of energy harnessed and controlled per capita per year by means of the agricultural and pastoral arts' (White 1949: 372).

They may be correct, but control over nature, and over animals' reproduction, lives, and deaths, denotes the emergence of civilisation metaphysically as well as physically. Each commentator directly equates human culture with power and control, in the latter cases reified as 'energy', and these are utterly orthodox views. Indeed much of the paraphernalia of western culture is orientated towards the accomplishment, demonstration, and eulogisation of such control. Science provides the expertise. Technological innovations, which we habitually regard as benefits in their own right, are tools for that task and intensification of production can almost be seen as an associated benefit. As Tannahill puts it, with domestication the 'farmyard animal became, in effect, humanity's first power tool' (1988: 27). The emphasis falls squarely on the word power as much as on the word tool.

In similar vein, Zeuner claims the presence of the dog to be a good index of human evolution in the late pre-agricultural period, being used by hunters in Africa, Australia, and the Americas. The Eskimo, for example, used the dog for hunting, transport, and food at times of famine (Zeuner 1963). The point at which we increased our power by using animals to control other animals indeed seems significant but the importance of such (anthropocentric) improvements is ideological as much as practical, demonstrative as well as enabling, affirmative just as it is effective. Domestication of animals, bringing them into the human fold as part of our stable of resources, serves as a signal of human superiority, much as the acquisition of a large and powerful car (the capacity of which is still measured in horsepower) does more than propel the modern business executive rapidly between two points.

As with the development of hunting, none of these views on domestication and farming can be literally accepted as true accounts of the development of *Homo sapiens,* since attempts to reconstruct human prehistory are inevitably conjectural. This does not, however, mean that theories about human evolution are not of interest; but they may be significant as much for what they say about contemporary definitions of what it is to be human, as for anything they tell of the past.

A more subtle analysis exists, however, than that of commentators who view hunting – like farming – as both the cause and the distinctive characteristic of changes in human organisation. It is equally possible to postulate that rapidly developing embryonic culture, perceiving itself in a new light to be different from, or superior to, the rest of nature, found in hunting an avenue through which to *express* that vanity. After all, an increase in meat intake was only one of a range of characteristics that are said to have appeared over the period, although we choose today to pay particular attention to that feature. Moreover, although many paleodieticians seem fixated by the development of hunting, other complex culinary and dietary discoveries – such as the storage, soaking, grinding, and boiling of seeds – may have been at least as significant in liberating humans from environmental constraints, enabling the species to colonise new habitats, and paving the way for agriculture (Washburn and Lancaster 1968: 295).

The apparent fact of so many contemporaneous changes in human attributes and activities is worth examining since different interpreters can come to very different conclusions after considering the same evidence. Sally Slocum, for example, disputes that prehistoric societies divided hunting and gathering between the genders on a pattern similar to that which is common today. She also takes issue with the assumption that this stimulated all the other known developments, including a

> longer gestation period; more difficult birth; neoteny, in that human infants are less well developed at birth; long period of infant dependency; absence of body hair; year-round sexual receptivity of females . . .; erect bipedalism; possession of a large and complex brain that makes possible the creation of elaborate symbolic system, languages, and cultures, and also results in most behavior being under cortical control; food sharing; and finally, living in families.
>
> Slocum (1982: 476)

Such interpretations, she suggests, may merely reflect the prejudices of modern, masculine society. As she points out, the word 'man' is commonly used ambiguously in discussions of hunting, supposedly to mean the human species, but actually being synonymous with males (1982: 474). For example, since Washburn and Lancaster imply that it is only males who hunt, and causally relate most human characteristics to hunting, Slocum suggests that the implication is clear, quoting Jane Kephart:

> Since only males hunt, and the psychology of the species was set by hunting, we are forced to conclude that females are scarcely human, that is, do not have built-in the basic psychology of the species: to kill and hunt and ultimately to kill others of the same species. The argument implies built-in aggression in human males, as well as the assumed passivity of human females and their exclusion from the mainstream of human development.
>
> (Kephart 1970: 5)

The evidence from anthropology, archaeology, psychology, and genetics, Slocum contends, simply does not support the male-biased speculative interpretations that are loaded on to it: skills such as co-ordination, endurance, and co-operation are not carried on the Y chromosome. It is nonsense with one breath to relate hunting prowess to such skills as control, leadership, and eloquence, and above all intelligence, and then to suggest that women are less well equipped to hunt than are men. Evidence that, on average, modern men are genetically better at judging distance and throwing accurately than are women (Kolakowski and Malina 1974) and that modern women on the whole have better vision in dim light and have sharper hearing than men (McGuinness 1976) does not prove that these characteristics determined their gender roles as is sometimes suggested (Tannahill 1988: 7–10). More highly developed abilities might just as well have evolved through competitive advantage in men's and women's socially prescribed functions.

In addition, Slocum argues, increasing demands for food would in most cases have been most efficiently met by extending gathering range than by increasing the time devoted to hunting. Discoveries of tools may be labelled as weapons, but might just as well have been used to aid in gathering and preparing vegetable foods. In short:

> By itself, hunting fails to explain any part of human evolution and fails to explain itself . . .
>
> To explain human nature as evolving from the desire of males to hunt and kill is to negate most of anthropology. Our species survived and adapted through the invention of *culture,* of which hunting is simply a part.
>
> (Slocum 1982: 483)

It is clear that more than one interpretation of the evidence is possible, and it seems likely that the interpretation we as a community have chosen, for what amounts to a pseudo-scientific origin myth, reflects what the dominant voices of our community have wished to believe of our past, in accordance with the preferred view of our present, as much as it reflects the reality of years gone by. It is significant, for example, that the human virtues supposedly promoted by hunting are those normally ascribed to the males of the species, in the western tradition at least. The recurrent stress upon such attributes as rationality, leadership, mastery, aggression, and male–male associations, and the consequently muted role implied for women, suggest that this entire theoretical edifice may merely rationalise the reality of the traditional man's world.

There is a long tradition of primitivism – the mythic image of people living elsewhere, or in times past, in a state of nature – as documented by Stephen Horigan (1988) who finds such descriptions dating back to antiquity. A revival occurred during the seventeenth and eighteenth centuries, in the wake of South Seas exploration (and at a time of industrialisation and urbanisation in Europe), and continues to this day; perhaps our archaeological interpretation partly reflects this enduring belief in prehistoric innocence. Horigan finds that the image of such people comprises several aspects, including various forms of a 'technological state of nature' and an 'economic state of nature', but invariably including a 'dietetic state of nature', namely vegetarianism:

> The latter form is particularly common; reference is often made in ancient literature to the 'milk drinkers of the north', and savage races who live on a diet of fish, milk, herbs, and berries. Indeed vegetarianism seems to have remained a popular element in the composition of such figures; mediaeval texts relating to the mythical 'wild man' often remark upon this. Homer, in the *Iliad*, refers to the land of 'the noble mare-milkers and the milk-drinking Abioi, the most righteous of men.'
>
> (Horigan 1988: 53)

If we do understand meat-eating to symbolise human distinction, then clearly pre-human 'noble savages' *must* have been vegetarian, and hunting must mark the emergence of civilisation.

It is in similar vein that adherents to the western world's major religion, Christianity, commonly ascribe the beginnings of meat eating to the Fall, when humanity lost its innocence and so became fully human. Our more recent quasi-religion – science – may, it seems, merely have translated the convictions of the old order into its own idiom and structure. Our intellectual establishment, dominated by male thought, has sought to authenticate itself by finding a historical rationale for its own perspective:

> 'I mean, it comes back down to the "ease" bit. I mean, why did man become a hunter? Because it was easier to go out and kill a boar, that might take half an hour or an hour's work, than to forage about for the whole bloody day picking up berries. It's a lot easier.'

Hunting is normally a less efficient means of obtaining nutrition than foraging, so why should we believe otherwise? Hunting is not necessarily less onerous, but it *is* more civilised in that it is demonstrative of human power over the hunted beast, and civilisation is known as a state of greater ease. It is thus more civilised – more human – to hunt wild animals than to stoop to forage all bloody day for berries. As civilised humans, we have long characterised ourselves as predators and conquerors. We continue to do so, and this notion of human power imbues our every channel of communication, including the food system. The notion of the Great Chain of Being survives, although today its earthly links have been scientifically reformulated as the Food Chain.

To believe that humans have no duties towards, or responsibility for, the non-human world, has the implicit consequence of legitimating meat-eating. Taken to its logical conclusion, this argument predicates that if animals have no rights, no feelings, no true independent existence, then there can be no sense in which it is morally suspect to use them for our own purposes as we see fit. If one accepts the premises without reservation, then cruelty is simply not a relevant concept. Thus, when the first efforts were made in England to obtain legal protection for members of other species (regarding bull-baiting), *The Times* newspaper thundered: 'Whatever meddles with the private personal disposition of man's time or property is tyranny' (25 April 1800).

This early bill was duly defeated. Even in the mid-nineteenth century, Pope Pius IX refused to permit the establishment of a society for the prevention of cruelty to animals in Rome, on the grounds that this would imply that human beings have duties towards animals (Turner 1964: 163). Although attitudes have since changed sufficiently to permit animals some protection from extreme abuse, agriculturalists still justify their slaughter and consumption on the grounds that to consider the interests of other species would infringe upon human rights:

> a large demand for meat exists in the UK, the EEC, and the world, and is in proportion to per capita income in most countries . . . To interfere with this demand would be highly controversial. It would be interfering with choice and freedom of action.
>
> (Wilson and Lawrence 1985: 25)

Freedom in this context stands for more than the right of the consumer to eat a particular foodstuff. As *The Times* leader hints by use of the term 'man', to restrict how the individual may treat other animals constitutes a symbolic – as well as a real – curb upon how humanity may dispose of the natural environment in general. Meat has long stood for the freedom to exploit freely. Julia Twigg notes that 'Meat was traditionally seen as the food of freemen and not of slaves, and beef in the eighteenth and nineteenth centuries was popularly regarded as the very basis of English liberties' (Twigg 1983: 23). And as Moscovici points out (1976: 1), 'individualism, together with the individualization of human actions, interest and relationships, tends to stress the contrast between nature and society [and the] individual is now the standard of reference in every sphere, be it physics, biology, economics or philosophy'. To dispute that the individual has unlimited rights over animals is to defy an almost sacred tenet of our common ideology – it is to imply that the power of human culture over nature is limited, and that is indeed controversial in a society in which human supremacy has for so long been a central ethos.

6

THE POWER OF MEAT

Belief in human dominion does not merely legitimate meat eating – the reverse is also true: meat reinforces that presumption. Killing, cooking, and eating other animals' flesh provides perhaps the ultimate authentication of human superiority over the rest of nature, with the spilling of their blood a vibrant motif. Thus, for individuals and societies to whom environmental control is an important value meat consumption is typically a key symbol. Meat has long stood for Man's proverbial 'muscle' over the natural world.

Sixteenth- and seventeenth-century moral theology, for example, seems almost directly to equate civilisation with the conquest of other creatures. Keith Thomas illustrates, for example, the sadistic vocabulary then attached to the important task of carving meat:

> Break that deer; . . . rear that goose; lift that swan; sauce that chicken; unbrace that mallard; unlace that cony; dismember that heron; display that crain; disfigure that peacock; unjoint that bittern; . . . mince that plover; . . . splay that bream; . . . tame that crab . . .
>
> (Thomas 1983: 25–27)

Even today, the slaughter of sentient creatures is not just a (perhaps regrettable) necessity in producing a valuable foodstuff. Bloodshed is central to meat's value. Indeed, the visible domination of other creatures is so important that cruelty is widely reputed to be *necessary* to producing high quality meat – veal production is one example. An advertisement protesting at South Korean treatment of 'pets', similarly, alleges that 'Kittens, cats and dogs will suffer appalling cruelty as they are slowly hanged . . .

strangled . . . clubbed . . . or tossed alive into boiling water. Terror stricken animals, it's claimed, taste better' (IFAW 1988: 3). In eighteenth century Britain too, Jennifer Stead notes that:

> Living fish were slashed to make the flesh contract. This was called 'crimping'. Eels were skinned alive, lobsters roasted alive, crammed poultry were sewn up in the guts, turkeys were suspended by the feet and bled to death by the mouth, bulls were baited before slaughter to make the meat more tender, pigs and calves were lashed for the same reason. One of William Kitchiner's recipes begins 'Take a red cock that is not too old and beat him to death'.
>
> (Stead 1985: 26)

Such practices, viewed as cruel from our perspective, were not simply wanton. Glorification of human dominance was then in its heyday so the harsh treatment of other animals is unsurprising. Meat was, and remains, a venerable symbol of potency, and indeed of civilisation itself. 'I do not believe', wrote a Swedish visitor, for example, 'that any Englishman who is his own master has ever eaten a dinner without meat' (Kalm 1748; quoted in Thomas 1983: 26). Even today meat is often linked with similar values: 'After nearly 12 years of negotiations, the tastiest symbols of American cultural imperialism are coming to Moscow in the form of McDonald's hamburgers' (*Guardian* 1987b: 6).

The humble hamburger is an appropriate, if somewhat clichéd, metaphor for North American culture since it embodies so many of the multi-faceted ideals on which the society is reputedly founded – of power and freedom, efficiency, and ease. When the ubiquitous McDonald's eventually began business in the Soviet Union, the hamburger's symbolic relevance did not go unremarked: '"It's like the coming of civilisation to Moscow," enthused Mr Yuri Tereshchenko, licking his fingers as he finished off a Big Mac' (Reuter 1990: 20).

The global 'hamburger joint' is the apotheosis of standardised production-line catering – its product a fitting food for industrial society, pre-cooked and camouflaged by the bread roll that relieves the stark savagery of raw red meat. The cherished image of sterile order seeks to reassure the patron that its '100% pure' beef product is above suspicion. And so successful has this strategy been that rising red meat sales in this market have offset otherwise falling consumption. Like so much industrial production, the mass-produced hamburger effectively divorces consumption from its ecological context. Fast-flesh emporia entice the consumer with sanitised gratification; here everybody smiles, whilst health, welfare, and environmental implications are banished to another less seductive world. In this role the hamburger has become the routine exemplar of 'junk' food, and the regular target for protesters' complaints. But the core symbol remains the ground beef by which the 'quarterpounder' or 'halfpounder' is measured, and the vital, virile, potency which that apparently endows . . . ideas by which even Gandhi was once influenced:

> It began to grow on me that meat-eating was good, that it would make me strong and daring, and that, if the whole country took to meat-eating, the English could be overcome . . . We went in search of a lonely spot by the river, and there I saw, for the first time in my life – meat. There was baker's bread also. I relished neither. The goat's meat was as tough as leather. I simply could not eat it. I was sick and had to leave off eating. I had a very bad night afterwards. A horrible nightmare haunted me. Every time I dropped off to sleep it would seem as though a wild goat were bleating inside me, and I would jump up full of remorse . . . If my mother and father came to know of my having become a meat-eater, they would be deeply shocked. This knowledge was gnawing at my heart.
>
> (Gandhi 1949: 3–4)

Gandhi sought the power of technology which he saw ruthlessly harnessed to subdue his people, just as it is otherwise brought to bear on the natural world. He thought that he might gain the desired force by physically consuming meat, mistaking the *medium* of expression of a more domineering tradition than his own with the *source* of its power.

Like Gandhi, and like many peoples who think that by consuming a physical substance one can somehow partake of its essence (see Chapter 12), we in the modern western world also seem to believe that meat alone can endow us with its unique vitality. We most favour the animal's muscle flesh since, metaphorically, we consume its strength. Unlike nutrition or crude economic logic, this helps explain our valuation of 'better' cuts of meat, for as Sahlins points out:

> The social value of steak or roast, as compared with tripe or tongue, would be difficult to defend. Moreover, steak remains the most expensive meat even though its absolute supply is much greater than that of tongue; there is much more steak to the cow than there is tongue.
>
> (Sahlins 1976: 176)

It is steak that constitutes the astronaut's ritual pre-flight meal (Twigg 1983: 23), affirming our technological control of the planet – technology on which the pilots' tenuous link is soon to be so dependent. Meat satisfies our bodies but it also feeds our minds. We eat not only the animal's flesh; with it we drain their lifeblood and so seize their strength. And it is not only that animal which we so utterly subjugate; consuming its flesh is a statement that we are the unquestioned masters of the world.

The motif of blood is central to the meat system. Indeed, it appears, by association with the colour red, blood is fundamental to much human thought. Berlin and Kay's cross-cultural enquiry into 'basic color terms' discovered that all languages have terms at least for black and for white (or dark and light), but that if a language contains three terms then the third will be for red. Red is the first true colour term to emerge as complex language develops (Berlin and Kay 1969).

Around the world red serves to suggest ideas of danger, violence, or revolution. Red is the colour of aggression, of power, of anger, of warning. It stands out and attracts our attention – hence its ubiquitous use in advertising and on fire alarms. As Edmund Leach notes of the oppositions we draw between red and other colours:

> When we make paired oppositions of this kind, red is consistently given the same value, it is treated as a danger sign: hot taps, live electric wires, debit entries on account books, stop signs on roads and railways. This is a pattern which turns up in many other cultures besides our own and in these other cases there is often a quite explicit recognition that the 'danger' of red derives from its 'natural' association with blood.
>
> (Leach 1974: 22)

Blood's mortal significance is indeed widespread. Blood is the stream of life itself – perhaps since life ends when too much is spilled. We faint at the sight of it. To have blood on our hands implies guilt – Lady Macbeth found a spot of it impossible to remove as she was driven mad. It signifies kinship: as in blood brotherhood, noble blood, or blood feud. It is used as the arbiter of inheritance as when we talk of blood lineage or say that some attribute is in the blood. The British upper classes are termed blue-blooded, as if so civilised as to be beyond mortality, no longer characterised by that most natural sign of red blood. Red meat might be suitable food for such beings, but it is apparently an inappropriate conception of their own physiology.

At Christian communion we drink the blood of Christ to form a mystical bond in which we partake of the Holy Spirit. In our horror stories characters such as Dracula drink the blood of living people to drain them of their life force. Blood is also the source of our passion – to be hot blooded is to be wild, spirited, lusty, impulsive; to be cold blooded is to be cruel, calculating, inhuman. The concept of lifeblood is evident throughout our culture, either directly or through association with the colour red. It is the so-called red meats, in which the blood is most vividly evident, which have traditionally been held in highest esteem in western society. It is red meats which today are most regularly reviled as unhealthy. It is also red meats which have been most zealously rejected by vegetarians.

Significantly, the image of red-coloured fruit and vegetables such as tomatoes or red apples seems to be largely unaffected by this association, possibly since by being categorically opposed to meat in the food system the link between their pigment and that of blood carries little meaning. Liquid red wine can, however, have something of the same reputation, upon which some 'full-bodied' brands such as Bull's Blood deliberately play, and as is made explicit in Christian communion. It is through red meat, however, that this gory image is most tellingly expressed:

> The concept of blood as the river of life continues to exert a hold even today. It is a force that no vegetarian should underestimate. It surely underlies the general disinclination of the average housewife to provide her family with soya protein instead of chunks of stewing: no amount of arguing about the nutritional sufficiency of soya products, it seems, can overcome this residual unconscious belief in blood . . . Vegetarians, therefore, are people who have somehow

> conquered this mythopoeic belief in the regenerative power of blood . . . Blood is the very stuff of life and meat partakes of its qualities and of its mythical and psychological associations. Sometimes no amount of factual evidence or moral exhortations can conquer this primordial logic.
>
> (Cox and Crockett 1979: 18–19)

Meat remains a graphic vehicle through which notions of natural human power are widely conveyed, and the image of blood is central to its efficacy. Cross-references between meat eating and such attributes as civilisation, instinct, ease and convenience, prestige, affluence, mental and physical well-being, potency, and skill, are routine, indeed endemic, in our belief, thought, and action, a common thread being the principle of environmental control and the benefits with which that civilised status reputedly endows us.

The remainder of this chapter will describe some ways in which the value of human omnipotence, sometimes expressed in terms of civilised standards, pervades different stages of meat provision and consumption: namely hunting; farming and marketing; and cooking and processing.

HUNTING

> Businessmen are to be taught how to trap, skin and cook wild animals on a four day survival course in West Perthshire . . . designed to boost their powers of leadership and initiative, and make them more self-confident.
>
> (*Daily Telegraph* 1988)

Hunting fulfils several functions: it can feed us . . . reduce competition for food . . . dispatch dangerous animals . . . it is even argued that the pursuit trains participants in the art of warfare. These are the practical reasons. But only its function of affirming our sense of prowess can adequately illuminate the cultural fact of contemporary recreational hunting. Since many societies subsist without the desire to hunt it seems unlikely to be a biological imperative, despite some writers' misconceptions. Washburn and Lancaster, for example, deduce from the millions of US tax dollars annually spent to supply game, that 'Men enjoy hunting and killing'; (unfortunately their lazy terminology obscures whether this trait is 'naturally' restricted to the male line):

> Many people dislike the notion that man is naturally aggressive and that he naturally enjoys the destruction of other creatures. Yet we all know people who use the lightest fishing tackle to prolong the fish's futile struggle, in order to maximize the personal sense of mastery and skill.
>
> (Washburn and Lancaster 1968: 299)

It is hard to understand how they can argue this case, whilst killing for 'sport' is rare amongst tribal peoples: Ingold, for example, describes how amongst circumboreal reindeer-hunters:

> one common problem seems to worry all the peoples with whom we are concerned. It is that whilst life depends on the harmonious integration of the various components or levels of being, this can only be achieved at one locus by breaking things up at another. Thus, the hunter lives by killing and eating animals, which inevitably entails their dismemberment. Much of the ritual . . . is designed to assist the reconstitution of the animals from the pieces into which they have been broken for the purposes of consumption, thus ensuring the regeneration of that on which human life depends . . . Above all, nothing should be wasted, for this would indicate a casually destructive attitude to nature which would only offend the animal guardians.
>
> (Ingold 1986: 246–247).

Although economic opportunism has recently fostered deer farming, 'sportsmen' have generally targeted only those species that are not normally domesticated, or even eaten. We rarely eat carnivores, for example, but this by no means implies that they should not be killed – far from it. The walls of many an ancestral pile, or dilapidated country hotel, display the big game trophies which hang in perpetual testimony to the courage and skill (and wealth and status) of the conqueror – or by association of the building's current occupiers. Here are exhibited the corpses of those creatures which are not farmed; to stuff and mount the head of a year-old lamb, a pig, or a cow would be regarded as peculiar to the point of tastelessness.

The prestige customarily reaped from slaughtering 'big game' is curiously related to its very non-utility. Ordinary meat might be prestigious relative to other foods, but it can be consumed, and its symbolic worth therefore transacted, by almost anyone. To bag a 'big cat', however, or an elephant, a bear, or a moose, cannot be confused with such mundane satisfaction. This is conspicuous non-consumption. Pursuing such beasts to the death is an unambiguous statement of lucre and leisure, and to this end these powerful and elusive beasts have been in a class of their own. The ultimate trophy to suspend from the wall for all to see as conquered is surely the lion. We invest upon it the accolade of King of the Beasts as if to say that even the most regal, the most powerful, the most awesome of creatures, is no match for us. And we lay the skins of fearsome lions, tigers, and bears on the floor as rugs, literally and symbolically to be walked all over.

It is entirely consistent with the centrality of meat as a symbol of environmental control that the hunting of defiant animals should have become a pursuit reserved largely for the privileged and powerful. Exploitation of natural resources has thus been structurally analogous to control of human resources. Fox hunting, for example, enables the few active participants ceremonially to display their sovereignty in pursuing a reputedly intelligent animal to its death. But that is not all; the hunt simultaneously vaunts supposed supremacy over the horse ridden, over the hounds commanded, over the servants or workers employed, and even over those without financial or social access to the pursuit (should they wish it). For, as Veblen notes of just this context, 'In order to gain and to hold the esteem of men it is not sufficient merely to possess wealth or power. The wealth or power must be put in evidence, for esteem is awarded only on evidence' (Veblen 1899: 36). Meanwhile the many camp-followers can ceremonially bathe in reflected glory. Their nominal participation in the rout of their prey is ritually affirmed through the custom of 'blooding' in which novice hunters are daubed with the blood of the captured quarry – the symbol of its life and spirit.

Participants in blood sports characteristically regard their activity as part of the natural order, just as social and economic privilege has been similarly taken for granted. Even in the Dark Ages the aristocracy of the remnants of the Western Empire were frequently more interested in hunting than in agriculture (Bowle 1979: 149), and Marion Shoard writes that so intense was William of Normandy's passion for hunting that 'most of Essex, Sussex, Surrey, Hampshire, large stretches of the north and west and parts of Scotland and Wales were declared royal forest', a special forest law effectively banning food production for mere commoners:

> It is impossible to establish the precise extent of the royal forests in William's day, but it is clear that almost a quarter of England was royal forest during the reign of Henry II in the mid-twelfth century; and that by the thirteenth century, after a period of decline, they still covered about one-fifth of the land surface of England.
>
> (Shoard 1987: 37–38)

Ever since, hunting in Britain has been a largely élitist pursuit, defended from poaching under threat of dire penalties, and limited to those with the money or connections required to gain access to the land:

> Gone, in most cases, are the owners or tenants who can still afford to stalk their ground purely for their own pleasure and to entertain their friends. Letting by the week or by the day is commonplace. The day tenant . . . has paid his £100–£150 for his day's stalking, not expensive compared to the £40 a day of 100 years ago, and he wants his money's worth.
>
> (Bowser 1986: 23)

Bloodsports may have maintained a wider social base in countries with a more egalitarian tradition than Britain – indeed in recent years the sheer scale of indiscriminate shooting of migratory birds in the Mediterranean region has raised the spectre of widespread extinctions. Even there, however, a similar ideology seems to obtain. Hunting remains an expression of human dominion, exercised and expressed through the media of individual skill and cultural technology.

In Britain the most populist form of hunting is fishing – it is indeed the nation's top participation sport. Again it is the element of skill that is most valued and the element of challenge that is most glorified. Describing trout fishing, for example, Max Hastings, editor of the *Daily Telegraph*, intriguingly combines the ethic of competitive mastery with a flourish of sexist imagery: 'I would rather go home empty-handed after a day playing a dry fly than catch monsters with a deep-sunk lure dressed like a saloon-bar slut' (*Independent* 1989: 16).

That fishing should be less exclusive an activity than hunting red-blooded land animals is appropriate since, in many respects, fish are conventionally regarded as only semi-animal: commonly less avoided by vegetarians and less esteemed by keen meat eaters. It is fitting that the defeat of a symbolically less powerful creature should also accrue less prestige. Only the largest and strongest of fish, such as the salmon in Scotland, or the shark in the Americas, rival land animals for the resistance they display and so for the perceived challenge to be met and overcome. Accordingly only these creatures endow comparable prestige. The largest are stuffed and mounted, just as the heads of game beasts are paraded on walls, exhibited in glass show-cases for all to pay homage – not so much to the mighty fish but rather to the captor whose name is immortalised on a discreet brass plate.

Human skill is extolled throughout the meat system, not least in the context of hunting, as another key value metaphorically representing our predestined prowess. This is not to dispute that highly developed skills may be admirable qualities in any context, nor that skilled craft can add economic value to a product. Its association with animal resource management, however, conveys more than the obvious. This can be illustrated by the case of deer hunting.

Enthusiasts are wont to present deer hunting, inspirationally, as a noble pursuit standing for elemental human tradition. Discourses on the topic are conventionally littered with references to the gamut of civilised virtues. Lord Dulverton, president of the British Deer Society, for example, writes:

> A little reminiscence of long ago – my first encounter with the red deer was when I was aged about three or four and living down at Dulverton. While being taken for a walk in an old oak wood we came upon 20 or 30 red deer stags among the trees with heads held high and on the alert, looking most impressive and beautiful. That was where my love of deer started so many years ago and has never deserted me . . . It was well over 50 years ago that I stalked and shot my first Highland stag and that I mention because there have been very many changes affecting deer and the stalking of them in the years which have elapsed. Our equipment is very different. In those days it was considered rather unsporting to use an optical or telescopic sight. Practically nobody did except the old and the infirm. Now it has been accepted that, provided we do not use the telescopic sight for taking too long or too risky shots, anything that helps towards a clean kill with the very first shot the better . . . There are still

> a few people who complain that it is downright cruel to shoot deer but we are trying to spread the message even among them that it is necessitous for deer populations to be kept under control for their own good and if they are not to do serious damage.
>
> (Dulverton 1986: 5)

Far from being cruel, human control exercised through deer hunting is portrayed as appreciating their beauty and in the animals' own interests – much as HRH The Duke of Edinburgh is reported to have said that 'grouse are in no danger whatsoever from those who shoot grouse' (Keating 1988: 17). It is as if we, as higher beings, must accept the responsibility of dominion, and order the world according to our design – so long as we approach the task with the same love and respect that we believe God to show unto ourselves.

> 'We are the conservationists,' insists Diana Scott, a joint Master of the Devon and Somerset. 'We hunt to preserve the deer, to keep the herds healthy; you have to be a countryman to understand our pride at being custodians of red deer, not destroyers . . . You have got to think of hunting as a replacement for the predator wolf when deer were really in the wild, not all that long ago comparatively'.
>
> (Keating 1988: 17)

The British Red Deer Commission has a remit 'to study both the conservation and the control of deer' (Dulverton 1986: 5). 'Control' is a term repeatedly encountered in such contexts, indicating the assumed correct relationship between humans and other creatures, and helping to explain why we continue to hunt for pleasure when it is no longer necessary for survival. Hunting reifies belief in the innate privilege of superior human skills. As Veblen long ago noted, hunters' adamant insistence that their motivation is a noble one, and their perpetual reluctance to acknowledge baser reasons, is intriguing in itself:

> Sportsmen – hunters and anglers – are more or less in a habit of assigning a love of nature, the need of recreation, and the like, as the incentives to their favourite passtime. These motives are no doubt frequently present and make up a part of the attractiveness of the sportsmen's life; but these can not be the chief incentives. These ostensible needs could be more readily and fully satisfied without the

> accompaniment of a systematic effort to take the life of those creatures that make up an essential feature of that 'nature' that is beloved by the sportsman. It is, indeed, the most noticeable effect of the sportsman's activity to keep nature in a state of chronic desolation by killing off all living things whose destruction he can compass.
>
> (Veblen 1899: 257)

Through hunting we demonstrate our ability to subdue the wild – normally by the use of cultural artifacts such as traps and rifles – and the kill is an integral component. Curious corroboration is provided by a 1988 British television advertisement promoting a snack bar called *Tracker*. A man in hunting attire is shown tracking a deer through tranquil wooded scenery. He looks intelligent, strong, tenacious, and purposeful. Presently the beast is squarely in his sights – when it is revealed that he aims to shoot his quarry with a camera, rather than with a gun. He then relaxes and enjoys his chewy cereal bar. The meaning is clear. This modern male prefers to appreciate nature's living beauty without having to injure it. A bar made from natural cereal, implicitly opposed to the meat which he is shown to shun by sparing the deer, is therefore his obvious choice.

The camera has largely superseded the rifle in modern big-game safaris since we awarded less kudos for the creatures' destruction, as some species became endangered and others extinct. In other blood sports, however, people still derive pleasure from their climactic mortal triumph over brute nature. The hunt remains 'a balance between the animal having the ability to use its instinct and physical ability to escape and the human being attempting, by skill and intelligence, to prevent this happening' (Marvin 1988: 132). Hunting is an affirmation of the superiority of our technology and civilised skills over the wilderness; it is not primarily a pastime whereby we commune as one with nature, nor whereby we feed ourselves, although that may be the preferred imagery projected. The venison from Victorian quarry, for instance:

> was of little value; much of it was consumed in the Lodge. I suspect the staff became heartily sick of it. Some would be given to friends. The local Doctor, the Minister, and some of the employees would get their share and, no doubt, the dogs lived well.
>
> (Bowser 1986: 23)

The concern expressed by Lord Dulverton, above, about the sportingness of high technology weaponry is particularly interesting, for deer stalking originated at a time when the longbow was the height of hunting technology. It is appropriate therefore that, with the increasing sophistication of equipment available, some should doubt whether the contest is still fair game, as also indicated by the secretary of a Deer Management Group, for whom challenge is the essence:

> The Victorians' idea of sport was such that many favoured the solid bullet, as it had to be placed more accurately to ensure that the beast was killed. Likewise, the telescopic sight was frowned upon. To quote Augustus Grimble, 'If the quarry will only keep still, it is apparently brought up almost within touch of the muzzle of the rifle and missing becomes nearly impossible. All the difficulties of judging distance, all the nicety of taking the sight in a bad light, all the pleasure in fact of making a brilliant shot with an ordinary rifle is done away with'.
>
> (Bowser 1986: 22)

When the element of skill required of the hunter is so diminished by the aid of technology that the animal has little chance in the contest it is proportionately less of a personal achievement to kill, so less prestige accrues. This must somehow be rationalised. In Michael Cimino's film of *The Deer Hunter* (1978), for example, Mike (De Niro's lead character) lays obsessive emphasis on hunting alone and reinstates the challenge by insisting on killing with one shot: 'Two is pussy . . . You have to think about one shot. One shot is what it's all about. One shot. I try to tell people that; they don't listen.'

Killing with a single shot is, for Mike (the rugged American loner), the only way to find real challenge. His partner Nick (Christopher Walken) responds by branding him a 'control freak'. In modern British deer-hunting idiom, conversely, the dilemma of increasing sophistication of weaponry reducing the scope for individual skill is resolved by stressing the social rather than the individual value of the hunt so that the technology is rendered acceptable rather than rejected. Whereas in the past hunting was principally an individual achievement with the assistance of cultural technology, today it has become increasingly a demonstration of technological power through the agency of society's hunters. Deer populations are controlled 'for their own good' and to reduce damage to human resources, so must be killed 'cleanly'. Sportsmanship is sublimated to social duty. Some insist that this is the only purpose, but others agree that the challenge is their ultimate delight:

> . . . in the last analysis, what is it that draws us, even when we are becoming feeble of foot and rheumatic of shoulder, to struggle up the hill once more to puff and pant round the tops, crawl through peat hags and bogs and having, one hopes, seen a nice Royal or two, perhaps even a thirteen pointer, to shoot that nasty old switch? Whatever the Antis, of whatever fringe may think, it is not the lust to kill – certainly not. It is because we still enjoy the challenge – masochists that we are – because we love the high places and first, foremost and above all, because we love the dun red deer.
>
> (Bowser 1986: 24)

FARMING AND MARKETING

Francis Klingender, analysing the treatment of animals in art and in thought up to the end of the Middle Ages, detects a basic conflict inherent in our approach to nature. He argues that the daily work of hunters, trappers, and fishermen; stockbreeders, cattlemen, shepherds, and butchers; rat-catchers and insect exterminators; trainers of horses, dogs, and other creatures that work for man . . . all these activities, and many more, represent one aspect of the relationship between men and animals. The companionable relationship with birds and beasts enjoyed by children and adults, the poet's delight in the song and movements of birds, or the beauty that artists perceive in animals, represent an opposing view:

> In these contrasting attitudes we encounter . . . a dilemma . . . rooted in the relentless struggle between man and beast . . . Although today we breed animals for meat, we still prey on them and the struggle continues, even if our hatreds are now chiefly confined to pests and microbes.
>
> (Klingender 1971: xxv)

Little, it seems, has changed, although describing domesticated livestock as 'prey' may make us uneasy. We are used to hearing farmers vocally defend their role as responsible custodians of the land. The soil, and its animal or vegetable stock, has traditionally been regarded as a raw material to be manipulated through the devices of human science, with the legitimate aim of maintaining the greatest possible control over all growth, to maximise the yield for the human population

A progressive tendency towards the industrialisation of agriculture is now expressed in monocultural systems where a single crop is maintained over vast tracts of land in the name of production efficiency, as if the landscape were a factory floor. Any floral or faunal intruders are designated pests and subdued by technological responses such as applications of chemical pesticides and, lately, manipulation of creatures' very genes. Modern agribusiness views its fields almost literally as battlefields, in which the enemy is any element not part of the human design, as illustrated by a chemical company's advertising:

> *MORE FOOD?*
>
> *Here's how Monsanto is pressing the attack*
>
> Technical research resembles a battle. The most effective attack is based on a survey of the situation, picks an objective, maps a strategy, then carries it out. In the world-wide battle to provide more food, Monsanto researchers have done just this . . .
>
> *Big Gun in the Battle*
>
> Most recently, Monsanto's attack on weed-losses put a heavy gun in the field. This molecule is N-(phosphonomethyl)glycine. Unlike the arsenal of selective herbicides – this compound is totally phytotoxic. It annihilates to the roots . . .
>
> (Monsanto 1976: 124)

Exponents of this view typically see warfare with nature as an honourable duty, and final victory as an achievable objective. Science and technology – as manifestations of human intelligence – are the keys to total conquest. It is in this spirit that intensive livestock rearing systems have been developed. In these engineered environments all variables are carefully monitored, including ambient temperature, the creatures' nutrient intake, and even genetic inheritance. Drugs and surgical procedures such as castration and de-beaking are administered to minimise natural variability and to maximise profitable traits such as rapid growth. Every aspect of the animals' lives and deaths are monitored, and system-administered. Meanwhile some farmers denigrate less meticulous systems of production for their lack of responsibility:

> 'I disagree sometimes with the way that fishing goes on because here we are actually farming something: we're working on basic numbers. We have a breeding unit. But people that fish in the seas . . . The sea is basically just being plundered, and no one is really putting any of that money back into the system, whereas on our farming systems we spend back into the unit. It's not farming. There it's just – the biggest boat, the smallest nets you can get etcetera, and just pull out as much from the sea as possible. And they don't give two hoots about it.'

An image of our stewardship of nature lies at the heart of the farming ethic. This conditions our view of the duty to manage the entire natural system. Other than amongst a few 'romantics', any land not 'improved' to reflect human designs is still widely regarded as worthless, just as wild animals were until recently hardly valued other than as gun targets. It is their breeding that adds value, as another vehicle for the exhibition of human skill:

> 'I had a passionate argument, with a farmer who was arguing that they had brought up the standard of meat. According to him if there were no meat eaters there would be no reason to keep those standards up. That's what they always say about horse racing too: that if there was no gambling they wouldn't have the tremendous bloodstock that they have.'

This anthropocentric and circular argument implies that animals are bred for their own benefit, much as deer hunting is held to be for the animals' own good, when in fact the improvements are clearly assessed entirely from the point of view of usefulness to humans. It is of scant use to a turkey, overloaded by copious quantities of body flesh, that it can reproduce only by artificial insemination and can hardly stand upright. The Belgian Blue cow, able to give birth only by Caesarian section, and the transgenic laboratory mouse, programmed to develop cancer within 90 days of life, are also clearly reared to our specifications alone. Nor is it greatly advantageous to a thoroughbred racehorse to be able to sprint along neatly flattened race-courses when in any natural setting its fragile limbs would be liable to fracture. The advantages wrought through skilled breeding lie in the animals' adaptation not to any natural ecological niche (wild turkey, wild horse) but to the categories – the economic niches – to which our culture assigns them (battery turkey, racehorse). We arrogate to ourselves divine power not only of life and death but of evolution and destiny itself:

> In the eighteenth century it was widely urged that domestication was *good* for animals; it civilized them and increased their numbers; 'we multiply life, sensation and enjoyment'. Cows and sheep were better off in man's care than left to the mercy of wild predators. To butcher them for meat might seem cruel, said Thomas Robinson in 1709, but, 'when more closely enquired into,' it proved 'a kindness, rather than cruelty'; their despatch was quick and they were spared the sufferings of old age.
>
> (Thomas 1983: 20–21)

The latest manifestation of this drive towards the so-called improvement of nature is the appropriately named science of bioengineering which seeks finally to wrest control over the form and future direction of the planet's ecology from the assumedly haphazard ordinance of mutation and natural selection in order to engineer the technologist's blueprint more efficiently. In classic Cartesian tradition the biotechnologist relegates sentient creatures to the status of mere machines. Only the terminology has been updated to the sophisticated twentieth century so that the analogy is no longer the springs and mechanics of clocks as described by Descartes. Now we have gone digital. The jargon of animals' 'genetic programming' conjures up images of computers which, despite their deceptive ability to mimic intelligence, in truth are entirely inanimate. As usual we justify the quest for advance with recourse to economic criteria and ultimate advantage to human welfare; but as usual our performance of such technological feats simultaneously demonstrates our ability to control the uncontrolled. It seems we simply cannot tolerate the notion that we are not, in the end, absolute monarchs of all we survey.

Even whilst we have come to cherish the glories of that which survives of nature, biotechnology and microbiology are amongst the most competitive, profitable, and prestigious scientific pursuits of the late twentieth century. At one time it was the higher animals that most resisted our control, and threatened or challenged us, and so which were the highest status objects of consumption. But today we have achieved more or less total power over the activities of such creatures to the extent that we have wiped many from the face of the earth. Our power over insects, microbes, or viruses is, however, less complete, just as our mastery of the genetic make-up of those higher animals which survive is still only partial. It is these creatures, therefore, that today are represented either as the greatest threat to human health or as the greatest opportunity for what is called advance: namely further control. Research at such microscopic resolution is currently reputed to be a potentially vast source of financial profit. It is therefore towards the challenge of understanding genetic coding or microbial biology that we devote much research effort, and it is those who effect advances in these chemical, biological, or microbiological endeavours upon whom we endow high status.

Control over nature underpins these most sophisticated arenas of technology. But the principle of domination attends other stages in the meat supply process: such as when the animal comes to be dismembered. In a recent advertisement headline for 'Speed' butchery equipment, for example, savage imagery sustains the message:

> ONLY SPEED HAS THE
> STRENGTH TO CUT HUNDREDS
> OF BEASTS DOWN TO SIZE
> (*Butcher & Processor* 1987a: 20)

The appearance of slaughterhouses in the mass media is largely restricted to the occasional item concerning legal regulations, and even then the ugliness of death is only rarely depicted. We prefer to keep the gory procedure out of sight and out of mind, and meat well distanced from regrettable reality. We pay others to carry out the task from which most of us would shrink if confronted, banishing their activity to sites on the margins of our settlements. The low-status afforded to slaughterhouse operatives in the social pecking order reflects our common distaste for a process which we nonetheless mandate through consumer demand. We may indeed despise them for their lack of sensitivity, even as we purchase their produce in anonymously sterile packaging. For those directly involved, slaughter is commonly seen as a necessary aspect of the process of rearing animals, its inherent unpleasantness being partially sanitised by technological procedures:

> *Q. 'Do you ever go to slaughterhouses?'*
>
> 'Yes quite often. It's very clean, you know . . . very clinical. It's humanely killed, hoisted up and literally within minutes they have the skin and everything away from the animal. It soon becomes part of a routine job; it's a bit of a factory system.'

Only meat from animals that humans have slaughtered is regarded as edible. We tend to shun beasts that have died from natural causes, explaining this with reference to hygiene and enshrining the interdiction in formal legislation – no matter that there is little evidence to prove that properly cooked meat from animals killed, perhaps, by accident is necessarily unhealthier for us than that professionally killed. One farmer, for example, after losing several sheep in heavy snows, recalls:

> 'Well, we couldn't eat them obviously. I suppose we might have got something for them but in the end we just slung them into a pit. You wouldn't want them lying around for too long, you see, in case of disease.'
>
> *Q. 'But if you got to them more or less straight away, and they'd obviously been fairly well . . . refrigerated . . . after they died – why not eat the meat?'*
>
> 'Oh, no, you can't do that. It's just not healthy. You never know what's been up to with the meat if you haven't seen it done yourself.'

The nineteenth-century French researcher, Decroix, put this proscription to a somewhat unorthodox test, cooking the flesh of animals that had died in various ways (including a 'mad dog') and feeding it to people without telling them its source; he noted no subsequent ill-effects in his participants (cited in Renner 1944: 125). Our ambivalence must again be interpreted in the context of our relationship to the meat. It is necessary that we slaughter the animals destined for our tables ourselves – or rather that it is done by those we employ to undertake the task behind closed doors. But the death must be at the hands of a human. For were we to eat animals that had died other than under our control then by our own definition we would be scavengers and that is not our favoured self-image. We may be powerful hunters, we may be skilled farmers, or we may be ingenious biotechnological manipulators, but we are not mere scavengers – we are in charge of our own destiny.

The marketing of meat is essentially similar to that of any other commodity. There are merchandising problems peculiar to this particular product, such as the need to adjust the sales pitch for some customers' sensitivity. But the essential process of selling animal flesh is similar to that of selling wheat, hi-fi, or nuts and bolts. It is taken for granted in the trade press that the beast is little more than a machine to convert the raw material (feedstuffs) into the finished product (meat) minus, of course, maintenance requirements. The problem is to make a profit, as a meat company manager laments:

> 'Meat actually struggles to get 20 per cent [profit] second time around. You're probably talking about 25 or 30 if you're a first-hand retailer of meat. But if you are purchasing it in, because there are two margins, you're probably looking at about 20. If you're looking for 25 you then kill the interest of the consumer because you're asking for too high a price for it and the alternatives that are on view in the supermarket chain, in ready meals with a meat-based content, are substantial. And also the convenience. So it's always a balancing act. That's the problem. We're looking at it every which way to encourage our market forces to be right.'

The very orthodoxy of our meat suppliers' assumption that slaughtered creatures should be accorded barely more consideration than sacks of coal or microchips is remarkable. Our indifference conforms, of course, to our classification of the entire non-human world as existing at a qualitatively lower level than ourselves – the ideological chasm separating *us* from *them* here being expressed in the incontrovertible idiom of western economics. Its laws are portrayed as natural so that considerations such as ethics can be of only secondary significance. The same manager, for example, admits:

'My knowledge of vegetarianism is very limited but I think obviously the attitudes towards morality will come from a certain section of people who are intelligent enough to consider it, and who actually get themselves very much involved in all the major issues. And they'd like to live in as near to an ideal world as can be managed. Now you and I both know that when you get into a much rougher market place then quite possibly that utopian idealism can't be sustained for financial reasons as much as anything, and just by the laws of nature.'

The inherent ideology is clearly communicated without ado: the economic 'reality' is that (non-human) animals are resources just as are vegetables or minerals. The very orthodoxy of its expression in the superstitiously incontestable terms of economics means that to dispute the distinction between human and animal would be to assault one of the pillars of modern society, with consequent marginalisation of the argument. The meat market is predicated upon the assumption that humans are qualitatively superior: our right to enjoy the food of our choice unequivocally precedes whimsical notions of animals' interests. The supposedly objective 'laws' of supply and demand call the tune (just as they presumably did when we traded in slaves). The dollar's jurisdiction puts our sovereignty beyond dispute.

In keeping with the almost regal stature which the innate privilege to exact our pound of flesh in tribute from the animal kingdom bestows upon all humans, butchery outlets traditionally emphasise personal service and skilled craft in preparing their wares. The recent 'Q Guild' initiative to raise standards amongst Master Butchers evokes medieval authenticity by its appellation, emphasising the butcher's pedigree:

[A Gloucestershire butcher] makes 12 to 18 prosciutto hams a year and sells it in thin slices for £1.88 per qtr. 'The process needs a lot of patience, particularly during the curing', he said. 'There's a lot of love involved'.

(*Butcher & Processor* 1987c: 6)

'My customers need to know that the meat they are getting is absolutely top grade and that's what I give them. They know if they come to me then it's nothing but the best . . .

It's very important to give people that quality because otherwise my customers will feel cheated and then meat might not have that special ring for them any more. Meat's all about that something a bit special.'

Apart from the implicit virtue of traditional service, the standard repertoire of meat merchandising has normally improvised upon two themes. The first has been to evoke meat's naturalness, although this has sometimes come to little more than small sprigs of plastic parsley separating the dishes on the white tiles in the butcher's window. More recently, however, a concerted effort has gone into projecting and protecting this image. Posters abound of healthy, happy chickens or pigs, set against a background of green rolling landscapes. Perhaps not surprisingly, few official images relate the production-line conditions in which most such animals are really reared. Terms such as 'natural goodness' develop an infinitely preferable marketing mystique.

The second strand is less explicit, and deals in ideas of rivalry, violence, strength, moral fibre, and mastery. The ambiguous imagery in recent slogans used by the British Meat and Livestock Commission, for example – 'Nothing Competes With Meat'; 'Slam in the Lamb'; 'Lean on British Pork' – is telling. The pages of the trade press offer numerous similar examples testifying to the special potency that is basic to meat's social character. One front page of the *Meat Trades Journal* (1989e: 1), for example, depicts Oscar Clark, a meat inspector, bending steel bars with his teeth; another of his off-duty activities, we are informed, is pulling meat wagons with his teeth, all for charity. 'Red packs quite a punch' proclaims another headline, leading an article on how a redesigned shop-front stressing the traditional meat trade colour of red achieved the 'punchy' look required (*Butcher & Processor* 1989b: 19). And meat advertising routinely portrays competitive situations – one promotion shows a young boy in karate costume, posed in hostile fighting stance and expression, astride a plinth with a trophy at his feet – presumably testifying to the food's legendary strength-giving properties. Those charged with the marketing of meat continue to emphasise power, and implicitly environmental control. They, at least, clearly believe that these values remain persuasive.

COOKING AND PROCESSING

We do not generally eat animal flesh in its crude state. With the exception of periodic vogues for raw foods such as the Japanese sushi, or for certain culinary specialities such as *steak tartare*, flesh almost invariably undergoes a transformation before we allow it to enter our mouths: we cook it. The next chapter considers in more detail the range of foods we classify as edible raw, and those which must be cooked, but meat is predominantly of the latter type. This apparently mundane observation is of singular significance since every known society cooks at least some of its food, and we are the only species which does so. Routine and ritual cooking of food is one trait by which all human groups can be categorically distinguished from all other animals. Or, more accurately, in this way we *distinguish ourselves* from other animals. The dichotomy between us and them is our mental creation.

Time and again, around the world, myths deal with the origins of fire, and fire and cooking play a key role in ritual, signifying its elementary importance to humanity. Prometheus stole fire from the Gods. The South American Gê people stole fire and the skill of cooking from the Jaguar (Lévi-Strauss 1970: 66). The Chukchi of Siberia have strict rules about the generation and transfer of 'genuine fire', their fireboards are revered as family heirlooms, and both play an important part in the sacrifice of reindeer. Amongst their neighbours, the Koryak, fire 'signifies the source whence [domestic] reindeer originated' (Jochelson 1908: 87; Ingold 1986: 267–269). In Northern Canada it is reported that 'the Chipewyan distinguish themselves from animals and eskimo' by avoidance of raw food (Sharp 1981: 231). And Audrey Richards notes that the 'savage' is quite erroneously supposed 'under the guidance of some superior natural instinct denied to his civilised fellows' to eat his vegetables raw (1939: 1). Clearly the notional savage, whom we invented to stand in contrast to our own supposed civilisation, will be characterised by general contradiction of cultivated behaviour; if we distinguish ourselves from barbarity by the cooking of meat it is hardly surprising that savages should be presumed to consume not cooked meat but raw vegetables.

The importance of cooking has been noted by many writers. Guy-Gillet (1981) argues psychoanalytically that humans, through cookery, unconsciously act as intermediaries between three orders: cosmological, zoological, and cultural. Carleton Coon (1955: 63) suggests that cooking was the 'decisive factor in leading man from a primarily animal existence into one that was more fully human', as did Boswell before him (Hill 1964: iii.245, v.33n). Dando too, investigating the history of famine, states that control 'of fire was a great step in emancipating humans from constraints found in the physical environment. Humans are distinguished from other animals by their general preference for cooked food' (1980: 13).

But it was Lévi-Strauss who underscored the notion of cooking as the fundamental articulation of the distinction between nature and culture. He noted that every known society processes at least some of its food 'by cooking, which, it has never been sufficiently emphasised, is with language a truly universal form of activity' (1966: 937). The transformation (by fire) which 'universally brings about the cultural transformation of the raw' (1970: 142) is, for Lévi-Strauss, the most profound and privileged expression of the transformation from nature to culture – or in other words of the way in which human beings conceive themselves as different from the rest of the natural world.

Of course vegetables may be cooked as well as meat, but vegetables are also eaten raw – for example in salads. Meat, however, is almost invariably transformed from its natural state, if not by cooking then through processing such as drying, marinating, or pickling. The processing of bacon by salting or smoking may indeed contribute to its image as a lesser meat, since its subsequent cooking will mean that it has been doubly transformed by the time it is consumed. It is not sold dripping blood as is 'real' meat. Bacon would rarely form the centrepiece of an important meat dinner. It can, however, be served as part of the Great British Breakfast, at a meal when meat is otherwise absent in Britain (although even this custom is in decline). And, unlike most meats, it can be served to support other flesh or animal products, in a game casserole, with chicken, or in an omelette, for example. Bacon is a subsidiary meat, of which even some 'vegetarians' remain fond.

Sausages, cooked pies, chopped ham, corned beef, or pâtés, likewise, rarely enjoy the same prestige as a piece of proper meat. Except in cases of hardship, they are better suited to the day's secondary meal, or perhaps as an appetiser at the main event. These lower-status meats may be seen as more appropriate to children than to Real Men who need real meat. They can be sold by any grocers, rather than only by proper butchers. Such items provide us with animal flesh in accessible form, but much of meat's peculiar mystique is dissipated in the process.

Cooking and other techniques metamorphose meat from a corporeal substance to an artifact of our culinary culture. When still in its raw state, in the kitchen or on the butcher's slab, meat is not yet the stuff of mouthwatering delight. But as the vivid redness of blood becomes a less hostile shade of brown, so the flesh turns from distasteful to tasty:

> BETTY 'It's funny though. I can look at a lump of meat, like a chicken, and do all the bits with it, and all the necessary stuff, and take out its entrails and what have you . . .'
>
> STEVE 'I bet you couldn't pluck it though!'
>
> BETTY 'Oh, I think I could because when I'm looking at it like that I'm looking at it, like . . . differently. . .'
>
> STEVE 'Something to eat . . .'
>
> BETTY 'No, that's the thing! I'm not looking at it like something to eat; I'm looking at it as this . . . thing here that I've got to do this with, in the same way that in biology you'd dissect . . . animals and rats and . . . you'd just do it. It's not something to eat until you've actually got it in the pan and you're cooking it and adding to it . . . and once it's beginning to cook. Then it becomes something to eat, but while it's just . . . a bit of animal lying on your chopping board it's just something that you've got to do.'
>
> 'I was visiting my cousin in London, and she can't have been completely vegetarian yet then but she certainly didn't normally eat much meat. But she knew that I did eat meat, so for some reason when I came to stay she thought she should cook it for me . . . So she bought this mince for us but then she was too squeamish to actually deal with it herself until it had gone brown! I had to fry it and stir it until all the red colour had gone – and then she took over and did the rest.'

Raw meat, dripping blood, is what is eaten by wild, carnivorous animals, not by civilised humans. We position ourselves above animals in general by eating meat, and above other carnivores by cooking it. Raw meat is bestial and cooking sets us apart.

Appropriately, the genre of horror fiction involving degenerate individuals is an area where we often encounter the image of humans eating raw flesh. For example, an infamous 'video nasty', outlawed by the British censors in the early 1980s, was a film entitled *Cannibal Holocaust*, the cover for which featured savage-looking women tearing with their teeth at raw, supposedly human, flesh – cleverly combining transgression of the cannibalism taboo with the added shock of seeing (female) humans consuming raw meat.

Many people today are reluctant to handle raw meat at all, reporting that whilst they do not mind eating it once cooked they find it difficult to deal with a substance in which the blood is still evident. Likewise, the smell of blood from raw meat is enough to dissuade many meat eaters from entering traditional butchers' shops, since superstores offer an attractively sanitised alternative. It is perhaps not surprising that the odour emanating from raw meat should find relative disfavour, although the smell is unpleasant in no absolute sense any more than the scent of a sizzling steak is automatically appetising. Either can be attractive or repellent according to our disposition. But smell can be highly evocative, capable of conjuring distant memories literally out of thin air. It is the associations which we find distasteful – the scent of bloody raw meat cueing us to consider the animal's death.

There are, however, occasions when the vulgar ubiquity of cooked meat is insufficient to communicate the desired message – when the potency of the symbol 'in the raw' perfectly conveys the severity of spirit implied. The stark barbarity represented by uncooked flesh has a rare capacity to disconcert us when its imagery is invoked in particular contexts. Raw meat, oozing blood, most strikingly represents the brute power of nature, undiminished by acculturation. In modern fields of combat, therefore, the idea of raw meat as the very essence of brutal nature, red in tooth and claw, can be an effective statement of extreme ruthlessness:

> Lloyd Honeyghan, looking every inch the magazine-cover picture of a world champion, returned to London yesterday . . . to announce the next defence of his WBC and IBF welterweight [boxing] titles . . .
>
> Honeyghan, resplendent in a £1,500 leather suit complete with studs, buckles, chains and horsehair epaulets

> – 'they're the scalps of my last two opponents' – will fight the WBC's No 1 contender . . .
>
> It will also be Honeyghan's second title defence in two months, but he declared: 'I want to fight as often as possible. I'm still so hungry for success I've been living on raw steak.'
>
> (Massarik 1987: 28)

> 'Oh yes, I know it's something that some businessmen do quite deliberately. You take the guy you're negotiating with to a fancy restaurant for a business lunch and then order steak tartare. It totally unnerves the other guy seeing you eating this raw meat with blood dripping out of it, and actually does make a difference – it can just give you that edge.'

Processing such as cooking transforms meat from a natural substance to a cultural artifact. Thus, the more skilfully its manipulation is effected, the better it expresses the supremacy of human civilisation. Cooked meat places us above the mere animal, and its appreciation affirms to us our privileged status, to which can be attributed at least part of its added value. Skill is for this reason expected of farmers, butchers, and chefs. The greater is the skill required in the processing the higher is the value of the end-product, and those we entrust with meat's provision must discharge their responsibility with diligence. In restaurants the chef's skill is paramount, since this is the arena of the specialist. Beef steak – cow's muscle – remains the most popular choice of entrée in British restaurants, and its preparation is a matter of grave concern, curious though this may seem to a vegetarian:

> 'I don't know if I'm just being really biased but I went out for a meal yesterday with my dad and my brother and his girlfriend and they all had meat and I had an omelette because they didn't have any veggie stuff on the menu and none of them liked their food! I keep noticing that whenever I go out with people and they eat meat and I don't and they don't like their food and I do. I've decided that either people don't know how to cook meat or people who eat meat are really fussy! It's either too tough or it's not cooked enough . . . and meanwhile I'm always quite happily munching into my pasta with tomato sauce or whatever I'm

having and there's never anything wrong with that. People just never seem to be satisfied with it. Maybe they expect too much: like they order a big juicy steak, and it's not big and it's not juicy. It's just so important to people that it should be just as they want it and it never is.'

The entire process of procuring and preparing meat bears evidence of a relationship between powerful, predatory, 'civilised', humans and our 'legitimate resource' of non-civilised animals. The proper texture for meat presented at table, for example, is a matter of fine discrimination. If cooking, which tends to make meat more tender to the tooth, represents to us the qualities of humanity, it is perhaps not surprising that tenderness should be a quality highly valued in cooked meat:

'And I did sirloin steaks in wine. Then again, that was in the casserole, because I find steaks can be a bit . . . no matter how good your steaks are, grilling them can be a bit tricky, you know. They can be tough, which, you know, can be a bit upsetting if you produce a tough steak to a guest.'

On the other hand, meat should certainly have some 'bite' to it: something to get one's teeth into, that puts up a bit of resistance – a quality with which the value of challenge in hunting curiously reverberates. Of all foods, only meat is held to have this proper texture that gives full eating satisfaction which is why, for many, meatless meals are incomplete.

If, as an anthropology textbook claims, central 'to every culture is its way of obtaining food' then strictly speaking we are no longer gatherers, nor hunters, nor even farmers, but we live in an age of industrial food provision. We 'go to the supermarket to choose among thousands of products marketed mostly by large corporations. Increasingly, these corporations control every step of the process of food production and preparation from the farmer's field to the fast-food restaurant' (McElroy and Townsend 1985: 175). And in this world, we are repeatedly told, meat is the ideal convenience food.

But convenience is more than a rational objective. Convenience is a by-word for the civilised society in which we have elevated ourselves above the daily grind of days gone by. In the words of the television advertisement: '*Menu Masters* help you make time to live your life'. We are masters. No longer do we merely save time – we make it, God-like. Convenience is leisure; convenience is the power to have the work done by other means; convenience is to be on top of the heap – and meat signals convenience. Why? Surely not for its functional attributes, since a slice of bread or a tin of baked beans would be equally convenient, nutritional, and filling. Meat is called a convenience food because it already stands as an expression of those same core values of modern western society: of power, of superiority . . . of civilisation.

7

THE BARBARITY OF MEAT

She looked down at her own half-eaten steak and suddenly saw it as a hunk of muscle. Blood red. Part of a real cow that once moved and ate and was killed, knocked on the head as it stood in a queue like someone waiting for a streetcar. Of course everyone knew that. But most of the time you never thought about it. In the supermarket they had it all pre-packaged in cellophane, with name-labels and price-labels stuck on it, and it was just like buying a jar of peanut-butter or a can of beans, and even when you went into a butcher shop they wrapped it up so efficiently and quickly that it was made clean, official. But now it was suddenly there in front of her with no intervening paper, it was flesh and blood, rare, and she was devouring it. Gorging herself on it.

She set down her knife and fork. She felt that she had turned rather pale, and hoped that Peter wouldn't notice. 'This is ridiculous,' she lectured herself. 'Everyone eats cows, it's natural; you have to eat to stay alive, meat is good for you, it has lots of proteins and minerals.' She picked up her fork, speared a piece, lifted it, and set it down again.

Peter raised his head, smiling. 'Christ I was hungry,' he said, 'I sure was glad to get that steak inside. A good meal always makes you feel a little more human.'

She nodded, and smiled back limply. He shifted his glance to her platter. 'What's the matter, darling? You aren't finished.'

'No,' she said, 'I don't seem to be hungry any more.'

(Atwood 1969: 151–152)

If eating meat is indeed such an important statement of human power, it might seem strange that we are apparently becoming progressively more uncomfortable with reminders of its animal origins. Consumer attitudes today are in a state of flux, not least for this reason. Whereas once it was sufficient simply to display whole animals and pieces of meat, the packaging of the product is now a more delicate task. Most of us prefer not to think too directly about where our meat has come from, and unwelcome reminders can be distinctly off-putting:

> 'I don't like it when you see bit of veins and things coming out of the meat . . . I think because it always reminds me of my own insides in a funny sort of a way. I suppose it's the idea of, like, blood flowing makes you realise that this slab of meat was once a bit of a functioning body, a bit like your own.'

Meat marketing has responded accordingly, to assuage customers' sensitivity to the nature of the product. Nowadays, the consumer need never encounter animal flesh in its vulgar undressed state. Instead it will come cooked and reshaped, in a sesame bun or an exotically-flavoured sauce, as a turkey roll or as chicken nuggets, in a crumb coating or a vacuum-package, with not a hint of blood in sight. More and more butchers' windows sport fresh green vegetables, fragrant herbs, and perhaps a stir-fry mixture. The most innovative diversify with in-store bakeries or specialist groceries. A deliberate process of disguising the source of animal foods has gathered pace in the twentieth century, reacting to our evident unease with the idea of eating dead animals:

> 'As a butcher I deplore deliveries being carried into my shop from the high street on the neck of a van driver – especially if they are not wrapped . . . I can think of little more guaranteed to turn pedestrians off buying meat than the sight of pigs' heads flopping about as he struggles past them with the carcase', said Mr van der Laan.
>
> Meat's connections with live animals had to be camouflaged.
>
> (Stewart 1989: 7)

Traditional retailing centres around offering the public bits of animals and often identifies meat with livestock. But

> modern consumer attitudes shy away from this link and so the butcher would be much better served by thinking away from the animal and more towards the meal when dressing his window and presenting his products . . . There is an urgent need for a new retailing philosophy. We are no longer in the business of selling pieces of carcase meat. We must make our customers think forward to what they will eat rather than backwards to the animal in the field.
>
> (*British Meat* 1987b: 4)

Independent high street butcher's shops have declined considerably in recent years. Whilst concentration has occurred throughout retailing, supermarkets have clearly derived particular competitive advantage from presenting meat in conspicuously hygienic conditions with all preparation completed out of sight. Often only the best cuts are displayed – bones, guts, and skin are nowhere to be seen. The hermetically sealed package is effectively dissociated from the animal to which its contents once belonged, a service that is clearly winning customers:

> 'No I don't often go [to the butcher's]. I'd as soon pick up a chicken leg or something from the Co-op when I'm going there anyway. I know that you don't get such good choice . . . it's just I've never liked these places you see . . . oh it just has that effect on my stomach . . . and then sometimes you see them with their bodies hanging there and that . . .'

The extent to which even the less radical members of society today perceive something distasteful about raw meat is illustrated by a 1989 television advertisement for Wall's pre-cooked sausages for the microwave. This segment of popular culture features an upwardly-mobile young man (apparent from the smart house and working-class London accent) taking the opportunity of his health-conscious wife's temporary absence to indulge in an old-fashioned treat updated. The characterisation of meat as naughty-but-nice is new and the emphasis on 'brown' – which occurs six times in the short slot – is significant, emphasising that this product is not blood-red like other meat. Talking to camera he hushes his audience:

> 'Shhh! Lucy, my beloved, seems to have over-oxygenated at her aerobics class so she's decided to have 20 minutes under the sun-lamp browning. These days everything has to be brown . . . brown rice . . . brown bread . . .

Well I'm keeping up with the Browns too with these: new Wall's sausages for the microwave. [Eats one cold, winks, and smiles]. They're pre-cooked so a couple of minutes in the micro and they come out piping hot and perfectly brown.

And they taste just like good old bangers. No mess . . . [glances as if listening for his wife upstairs] . . . and no evidence that anything but celery hearts and nut cutlets were ever here. Now that's what I call nouvelle cuisine!'

The names we give to the flesh of the main meat animals are another device whereby we reduce the unpleasant impact of having to acknowledge their identity. We do not eat cow, we eat beef; we do not eat pig, we eat pork; we do not eat deer, we eat venison. It is as if we cannot bear to utter the name of the beast whose death we have ordained. According to Edwin Ardener:

> Sir Walter Scott drew the conclusion that the split in the English categories reflected the fact that the English knew the product on the hoof, whereas the Normans received it cooked. The perpetuation of the division when the Normans and English became one speech community is less easily explained.
>
> (Ardener 1971: xxix)

Our continuing use of foreign terminology instead of the more direct Anglo-Saxon is, on the contrary, readily explained by our desire to avoid the full conceptual impression of stating the name of the devoured animal – as part of a pattern of such mental manoeuvres. Why the British should be less inclined than the French to confront the reality of our repast is another question, but may merely reflect a mutation of long-standing national differences in culinary philosophies:

> A Norman would have despised the barbarous magnificence of an entertainment consisting of kine [cows] and sheep roasted whole, of goat's flesh and deer's flesh seethed in the skins of the animals themselves; for the Normans piqued themselves on the quality rather than the quantity of their food, and eating rather delicately than largely, ridiculed the coarser tastes of the Britons.
>
> (Scott 1829: 11)

In contrast to our medieval forebears, the modern British are reluctant to receive unessential reminders. The term fleshers has gone out of fashion and there are signs that even the word butcher, with its associations of blood-stained savagery, may be going the same way. Refurbished outlets increasingly bears such names as 'meat market' or 'purveyor of fine meats' which again stress the food itself instead of its bestial source or its dismemberment. Likewise, the place where we have our animals put to death is nowadays less often officially referred to by the graphically descriptive term of slaughterhouse. The French term abattoir nicely serves to ameliorate the brutal reality of their service to our society, and the British *Meat Trades Journal* recently suggested the Americanisms 'meat plant' and 'meat factory' to further divorce the image of the food from the act of slaughter (Serpell 1986: 158–9).

> Verbal concealment crops up in most areas of animal exploitation. We speak of the 'harvest of the seas' as if fish and shellfish were analogous to wheat and barley, and in the fur trade 'pelts' are 'harvested' rather than animals flayed. It is even tempting to suggest that much of the technical jargon employed by scientists who experiment on animals is simply an elaborate euphemism, a method of disguising the animal's affinity with humans, and so promote detachment.
>
> (Serpell 1986: 159)

To some, our willingness to consume meat as well as the many other assorted products of the animal industry, but apparent unwillingness to slaughter the beasts for ourselves, or even to acknowledge our complicity in that process, is a matter for moral reproof:

> 'I think in Britain the meat industry is very dishonest. The people are not allowed to be aware of what's going on. To them meat is wrapped up in cellophane in supermarkets; it's very divorced from the animal that it's come from. The thing is that people see these intensive farming places as being unpleasant but they avoid taking any personal responsibility for it and so everybody just accepts this thing – that eating meat is okay so the intensive conditions are okay. But it doesn't feel alright to me. People don't go down on the factory farm to see what's really going on down

> there. I think if a lot of people did do that or go to the slaughterhouse to see how the meat is produced then a lot of them would become vegetarians.'

There is some evidence to support this belief. Many first-generation vegetarians or semi-vegetarians directly trace their abstinence to occasions when, for one reason or another, they were brought face to face with the connection between the meat on their plate and once-living animals. The particular incident related by any individual – be it the sight of carcases being carried into a butcher's shop, or an encounter with vegetarian polemicism, or a visit to a slaughterhouse on daily business, or merely an unusually vivid flight of imagination – is of minor importance. What matters is that many people, when confronted with this ethical perplexity, seemingly prefer to forgo meat altogether rather than to condone the treatment of animals on their path from birth to plate. And equally important, perhaps, is how new is this rebellion, or rather how rapid has been its development in recent history.

Norbert Elias documents the thrust of social change over time in his epic study of the history of western manners from the Middle Ages, *The Civilising Process* (1939). Through extensive illustrations, Elias argues that standards of 'civilised' behaviour have moved in a consistent direction. Two of his examples are particularly relevant here, the first being his biography of meat's ideal representation over the period:

> In the upper class of medieval society, the dead animal or large parts of it are often brought whole to the table. Not only whole fish and whole birds (sometimes with their feathers) but also whole rabbits, lambs, and quarters of veal appear on the table, not to mention the larger venison or the pigs and oxen roasted on the spit.
>
> The animal is carved on the table. This is why the books on manners repeat, up to the seventeenth and sometimes even the eighteenth century, how important it is for a well-bred man to be good at carving meat . . . Both carving and distributing the meat are particular honors. It usually falls to the master of the house or to distinguished guests whom he requests to perform the office.
>
> (Elias 1939: 118–119)

With time, however, reminders of the animal nature of the food are removed: first the head, or feet, or tail, and so on. It progressively becomes less polite for the joint to be brought to table, it instead being carved and served on a sideboard . . . and then even further removed, in the kitchen. Sauces, aspics or other presentational devices are increasingly used which effectively disguise the meat still further:

> The direction is quite clear. From a standard of feeling by which the sight and carving of a dead animal on the table are actually pleasurable, or at least not at all unpleasant, the development leads to another standard by which reminders that the meat dish has something to do with the killing of an animal are avoided to the utmost. In many of our meat dishes the animal form is so concealed and changed by the art of its preparation and carving that while eating one is scarcely reminded of its origin.
>
> (Elias 1939: 120)

Writing prior to the Second World War, Elias was not afraid to extrapolate into the future. Modern evidence supports his prescience, since general public standards have indeed been pushed yet further in the intervening years:

> There are even *des gens si délicats* . . . to whom the sight of butchers' shops with the bodies of dead animals is distasteful, and others who from more or less rationally disguised feelings of disgust refuse to eat meat altogether. But these are forward thrusts in the threshold of repugnance that go beyond the standard of civilized society in the twentieth century, and are therefore considered 'abnormal.' Nevertheless, it cannot be ignored that it was advances of this kind (if they coincided with the direction of social development in general) that led in the past to changes of standards, and that this particular advance in the threshold of repugnance is proceeding in the same direction that has been followed thus far.
>
> (Elias 1939: 120)

The second relevant trend noted by Elias concerns the well-mannered use of knives. Over the same period, he says, the knife has evolved considerably from its role as a sharp instrument for carving and transporting meat to the mouth:

> In the Middle Ages, with their upper class of warriors and the constant readiness of people of fight, and in keeping with the stage of affect control and the relatively lenient regulations imposed on drives, the prohibitions concerning knives are quite few. 'Do not clean your teeth with your knife' is a frequent demand. This is the chief prohibition, but it does indicate the direction of future restrictions on the implement. Moreover, the knife is by far the most important eating utensil. That it is lifted to the mouth is taken for granted.
>
> (Elias 1939: 122–123)

With time the knife evolves into an implement used only as a cutter, to its gradual bluntening as it becomes more of a pusher, to the modern American convention of its being used to dissect food at the beginning of a meal, thereafter being left completely alone. Discussing the prohibition on the knife being lifted to the mouth, Elias acknowledges that an element of rationally calculable danger is indeed present:

> But it is the general memory of and association with death and danger, it is the *symbolic* meaning of the instrument that leads, with the advancing internal pacification of society, to the preponderance of feelings of displeasure at the sight of it, and to the limitation and final exclusion of its use in society.
>
> (Elias 1939: 123)

An interesting development of this trend, half a century later, is the daily use of chopsticks in preference to conventional cutlery by many British vegetarians today. It is a curiously apposite corroboration of Elias's analysis that many of those individuals who profess themselves to be acutely concerned about violation of other human and non-human inhabitants of the world, and who express that concern through avoidance of animal flesh, should ultimately reject the knife altogether.

This is not to imply that these changes occurred steadily. There have of course been countless sub-themes and reactionary movements, times of little change and times of rapid change. The essence of Elias's argument is that the overall direction of change is constant. In the long term the tastes of one generation develop upon those going before, and the 'dice are loaded' in such a way as to result in a consistent pattern of progress: away from direct exultation in conquest towards more urbane values (1939: 118–119).

The trend towards avoiding reminders of the animal origins of meat can be seen as part of a wider movement in human behaviour, away from instinctive, biologically governed activity in favour of socially determined patterns. Hans Teuteberg notes, for example, that human sexual function is no longer cyclical, adding that our food habits, similarly, have been increasingly governed by choice rather than by nature:

> Throughout history, a gradual loss of instinct . . . is apparent . . . Apart from initial hunger, an artificially evocable appetite led to the surplus intake of nourishment not required in the calorific system of man. The initial adaptation of diet behaviour according to nature was changed in favour of socio-cultural factors . . . Although the loss of instinctive diet habits often leads to overfeeding, it contains the possibility of . . . enhancing culture itself.
>
> (Teuteberg 1986: 14)

A similar principle is evident in Eckstein's (1980) five-level development of the psychologist Maslow's 'Hierarchy of Human Needs' (1943) which he analyses in the context of food. At the most basic *survival* level, Eckstein suggests, humans will eat practically anything, such as in times of famine, crisis, or war. However, as Marshall Sahlins puts it, 'men do not merely "survive". They survive in a definite way' (1976: 168). Thus, Eckstein says, *safety and security; love and belongingness;* and *self-esteem* become valid considerations as each is satisfied, until *self-actualisation* is the key concern with foods classified according to a symbolism relating to personal identity. This rather coarse categorisation cannot be taken at face value, yet the communicative element of food selection may well have become more significant in the affluent modern world which, perhaps complacently, regards its food supply as bountiful and secure.

Elias is not alone in positing the gradual movement of culture away from purely competitive and hierarchical values. Todhunter suggests that this 'century may be called the era of social conscience – a concern for the welfare of fellow man' (1973: 287). Perhaps more significant still, however, is the new concern for non-fellow-man. As Lecky notes in his nineteenth-century *History of European Morals.*

> At one time the benevolent affections embrace merely the family, soon the circle expanding includes first a class, then a nation, then a coalition of nations, then all humanity, and finally its influence is felt in the dealings of man with the natural world.
>
> (cited in Lecky 1955: 100–1)

If there has indeed been such a direction in human development, then such values can truly be called 'urbane', since this trend is clearly related to our population's agglomeration in towns and cities:

> simple geographic evidence almost requires a link between militant kindness to animals and the New England of factories and cities . . . To note that city folk take a livelier interest in kindness to animals than do farming populations is no novelty; indeed, it is almost a commonplace. Clearly urbanisation and industrialisation in some way helped to generate the new concern for beasts. But this is merely an observation, not an explanation.
>
> (Turner 1964: 25)

Julia Twigg likewise finds urbanisation to have been a major motive force in stimulating a widespread reappraisal of the relationship between human society and other animals, and thus of meat as proper food:

> Traditionally the imagery of meat and blood was developed within the wider context of humankind's higher and lower natures. People's higher nature pertains to the rational, the spiritual and the moral spheres, whereas the lower relates to the bodily and to all that is designated their animal nature. This division into higher and lower has its counterpart in a profound series of oppositions between up and down, heaven and earth, and mind and guts/genitals.
>
> Though this is an image of great longevity and power and one whose influence still contributes significantly to the meaning of meat, it is today a model that less clearly holds sway; other perceptions concerning bodily existence and the meaning of the animal have come into cultural prominence. The most important of these result from the accelerating growth from the eighteenth century of tender mindedness towards animals . . . The causes of this major

> shift in consciousness are obscure, however one factor of undoubted significance is the growth in urbanisation.
>
> (Twigg 1983: 26)

One reason the movement of people to cities should have this effect, Twigg suggests, is that it breaks the organic contact of people and animals and throws into relief the arbitrary distinction between animals as food and animals as pets – a theme resumed in Chapter 9. It is little surprise that those brought up in rural communities, where the rearing and killing of animals is a normal part of life, should continue to accept such realities more readily than urban dwellers who have neither experience nor wish for experience of such things, or who believe that civilisation should transcend such customs. One woman, who became vegetarian upon moving to live in a city, recalls that until she lost the experience of daily contact with farm animals she had seen little reason for concern:

> 'I mean as children, in the cattle fields, we used to go and sit on the gates and look at them all and have favourites and ones that would come up and let us pat them. And we'd spend hours round with them all. And I suppose I've always liked animals but I can't say that I ever, as a child . . . I sort of thought it was horrible when they went away but I didn't really think about it. We were never . . . we never talked about it with my brothers and sisters or anything. I don't think it was a conscious thing. I didn't ever think "how awful that we breed them to eat". It was just a natural thing that I didn't really analyse like that.'

If control of nature is broadly speaking a cultural imperative then it is fitting that most major developments in culture should occur within urban settings. Towns are principally human space: created, defined, and maintained by people. The countryside is less human: less ordered and more subject to the vagaries of nature. Attempts to impose human control there are limited by inherently uncontrollable factors such as the weather. Thus, in the popular conception of town dwellers, country people – almost by definition – are less civilised since they are more directly associated with those natural processes that urban civilisation has sought to transcend. This is a characteristic which Gary Marvin, for example, observes of bullfight-hosting communities in Spain:

> This notion of human control and where, how and when it is exercised is a fundamental concern in this culture. To be fully civilized is to be in control of one's self, in control of one's life and in control of one's environment. This control is a function of will which people put into operation to overcome, on the personal level, their own human animal nature and, more generally, the world around them. Control is thus the domain of culture; lack of it signals the domain of nature. To be fully civilized is to be fully removed from nature, especially to be removed from the effects of its unpredictable elements. To be civilized is demonstrated by living in the urban realm with fellow human beings, by emphasizing that which is distinctively human (as opposed to animal) in terms of behaviour.
>
> (Marvin 1988: 130)

It is appropriate that whilst the ethos of human domination was in the ascendant as the ultimate expression of civilised values its vanguard activities, such as science and industry, should have been concentrated in urban centres. It is perhaps also appropriate that reaction *against* that ethos in expressions of compassion or in such traits as vegetarianism – the 'forward thrusts in the threshold of repugnance' that Elias suggests may be in the forefront of the civilising process – should also have been a largely urban phenomenon. If advances in social progress are indeed centred upon urban settlements, the relatively high incidence of vegetarianism in our urban centres of influence may provide a significant indication of the future direction of 'cultured' thought on the topic of animal foodstuffs.

The recent explosive growth of information and education might also be pointed to as an associated motive force for change, since awareness of the various issues and ideas involved with meat appears to be such an important variable. If there has been a movement in favour of avoiding reminders of meat's animal nature then the so-called 'information revolution', from which few are immune, might be expected to count against meat eating. We are bombarded with a diversity of images through countless media, some of which inevitably confront us with information we might otherwise prefer to avoid. A wider diversity of views is more widely communicated through different media today than at any time in the past – a potential threat to the status quo, as a representative of the Meat and Livestock Commission warns the industry:

> Children come home from school having heard opinions and participated in discussions about many aspects of life that would have found no place in school curricula 20 to 30 years ago. The established way of things is less readily accepted; alternatives are more readily explored . . .
>
> So the science of systematic exploitation of media opportunities is as well understood by small specialist groups as it is by large companies or industry organisations. However, there is the additional benefit to the former that the minority view is more interesting.
>
> (Harrington 1985: 2–3)

Diversity of views is not sufficient to *cause* change but increasing availability of information may be hastening the pace of change. Change also has a curious way of perpetuating itself. Market research shows that home cooks today choose their recipes by the photographs in cookbooks, and few such books today dare depict close-ups of bleeding sides of meat. Indeed, a cursory scan of most bookshops' shelves can almost give the impression of a nation of vegetarians, so prominent have been the positioning of 'healthy', 'natural', and meatless advice. Thus the traditional roast is caught up in a spiral of decline. The common thread to so many of the ideas discussed in this work, as in Elias's history of manners, is of human civilisation as a constantly developing process.

Recent advances in humane thought, particularly since about the time of the Industrial Revolution, can be viewed partly as a reaction to the excesses of the prevailing vainglorious ethos. The absolute distinction that we have tended to draw between humanity and the rest of nature is by no means universal or instinctive. Indeed it is doubtful whether the principle of nature subordinated to human demands has anywhere been so popular as amongst our scientific, industrial, and intellectual élites, upon whom it bestows considerable benefits and where it continues to flourish.

In medieval days, for example, the idea of humans coexisting with a living earth was more general (Merchant 1982: 1–41). Even in the eighteenth and nineteenth centuries the relationship is said to have been more dialectical for most people, with culture affecting nature and vice versa (Brown and Jordanova 1982). Animals were of course seen as different from humans, and normally inferior, yet were still commonly accredited with 'human' attributes such as reason and sensibility, much as many of us regard our pets today. Indeed, even into the nineteenth century, in Europe, bees, pigs, cocks, and even weevils were still accorded enough responsibility to be tried for crimes against human laws or against nature with full legal representation and sentencing (Evans 1987). But Cartesian dualism and mechanism held sway amongst our scientists and technocrats, and the western cultural diaspora, through conquest, trading, settlement, or imperialistic religious missions, ensured the dissemination of this view of nature to civilisations where such conceptions had been hitherto entirely alien. For example, the Sioux chief Luther Standing Bear recalls:

> We did not think of the great open plains, the beautiful rolling hills, and winding streams with tangled growth as wild. Only to the white man was nature a wilderness and only to him was the land infested with wild animals and savage people. To us it was tame . . . Not until the hairy man from the East came and with brutal frenzy heaped injustices upon us and the families we loved was it wild for us. When the very animals of the forest began fleeing from his approach, then it was for us the Wild West began.
>
> (Quoted in Brown 1972: 86)

Of course most peoples must control their surroundings to some extent, if only to fend off potentially dangerous predators, and so modify their ecology in the process. This does not, as such, say much about anyone's philosophy of nature. It is indeed wrong to assume that less industrially developed societies necessarily live in a more harmonious relationship with nature. There is little reason to romanticise such cultures, or our own past, since however we may characterise others in contrast to our own supremacist creed, modern western society is not alone in the damage it has inflicted upon its habitat (Hughes 1975; Harvey and Hallett 1977; Greene 1986):

> While particular types of industrial pollution may be new and the scale of ecological devastation may be greater now than previously, the modern world is *not* confronting

> completely unprecedented circumstances – numerous civilizations before our own have confronted environmental degradation and have paid the price . . .
>
> Recent archaeological research indicates that there was a substantial ecological component to the emergence and collapse of agricultural complexes in ancient Mesopotamia, Phoenicia, Palestine, Egypt, Greece and Rome . . .
>
> Over time the strategies that each society pursued to produce food and procure resources left their characteristic mark on the environment. Some of these strategies proved not to be sustainable and overtaxed the regional natural resource base resulting in the depletion of water, soil, or forest reserves. The general pattern was one of gradual emergence, brief flowering, and rapid collapse . . .
>
> (Weiskel 1989: 98)

Others may have wreaked comparable damage upon the world. But the unique aspect of the modern situation is the sheer extent to which control has finally been achieved and the degree of damage to complex natural systems which has been inflicted as a consequence, if not an objective, of the struggle. It is the unprecedented scale of our massed impact which leads many people today to question whether activities which might have been regarded as natural and thus acceptable when conducted on a smaller scale can necessarily be so considered when undertaken at the global level:

> 'I mean, people say that it's natural for us to kill animals and hunt and so on because we've always done so . . . but, well, beavers build dams and cows fart [laughs] but they don't shit concrete all over the whole bloody world like us or turn the atmosphere into one big chemical experiment.'

There is a growing perception that the need for a war of attrition against the natural world has passed, and that civilisation can mean more than the expression of overwhelming force. The battle has largely been 'won' even if the victory in the end turns out to be pyrrhic. As Dr Robert Runcie, Archbishop of Canterbury, put it: 'The initially endangered species, humankind, has become the endangering species' (Schwartz 1989: 2). Or, in the words of a food industry pundit, intriguingly arguing that environmental concern is a modern luxury, our 'ability to tame nature has caused today's generation, unfamiliar with her destructive capabilities, to view nature as only good, kind, safe, wholesome, and superior' (Chou 1979: 19). Chou evidently regards the non-specialist public as ignorant of nature's enduring menace. However, parts of that public evidently regard the high-technological food industry, both literally and metaphorically representing the entire industrial-scientific system, as a potentially greater threat to their health than its wild counterpart.

Today many dissident individuals and groups stress a counterposing view of culture. Instead of emphasising control and separateness, these heretics represent society as one special component of the entire living world, potentially living not by parasitic manipulation but in dynamic interaction with nature. Typical is a feeling of regret at loss of what is seen as our sense of unity with natural processes, and refusal to accept mechanistic dualistic doctrine as necessary or inevitable. This need not require the negation of culture through retreat into some mythical pre-industrial past; rather there is a common perception that our survival requires us positively to utilise our cultural attributes and advantages to develop a new state of more peaceful coexistence within the world. This distinction helps to resolve certain common misapprehensions. For example, many vegetarians and others seem happy to subscribe simultaneously to propositions which, on the face of it, can appear mutually contradictory – such as that humankind should return to a more natural way of life and also that we must evolve to a more civilised level of development. This only seems problematic, however, under the orthodox doctrine that civilisation and nature are intrinsically opposed, which is an assumption that many no longer accept.

Meat merchandising has long benefited from its product's embodiment of natural domination. Some people, however, choose to avoid it, also for that which it represents. Many of the ideas involved are held in common, but the degree to which these are *positively* received varies markedly. Refusal to eat meat conspicuously signifies rejection of more than the substance itself:

> 'I mean when I see people going into butchers' shops it really cuts me up – yeah almost literally – It feels like that . . . as if it's me that's on the block because I'm sort of responsible for it. Actually it really makes me cry sometimes. I just think, God, how can all these people go on gorging

> themselves on such lovely animals without caring at all? It's the whole society which encourages it and makes it seem like it's something wonderful to be so horrible and barbaric. I have visions of these people as monsters with blood running down their chins . . . but to most people it's just normal – that's what I find hard to understand.'

Red meat has traditionally been widely endorsed as a potent expression of sheer brute power, particularly for 'red-blooded' males. Today, meat still represents much the same principle, but for some the desirability of that trait may be redefined, and meat's attractiveness reduced accordingly:

> Meat can here stand not for maleness in an approved sense, but for what is seen as a false, macho stereotype of masculinity. Thus 'strength' and 'power' become 'cruelty' and 'aggression'; masculine vigour and courage become violence and the forces of human destructiveness. This perception is epitomised in the pacifist critique of war and of militaristic values, with which vegetarianism has close associations.
>
> (Twigg 1983: 27)

It is not necessarily the fact of meat which is rejected, in terms of health or nutrition, nor even the direct implication of how the particular animal in question has been treated and killed. For some a diminishing taste for meat may indicate only a vaguely formulated unease with aspects of modern animal management, which might be obliquely articulated as an expression of concern about meat's healthiness or price. Others explicitly revile the entire complex of cultural values which engender ill-treatment of the non-human world, and so reject the icon of meat as well as much of the output of the technological food processing industry.

For many non-meat eaters the central assumptions of the orthodox food system are essentially reversed. For example, red meat (which intrinsically implies the death of a highly evolved animal) has conventionally been our most prestigious food, roughly followed in terms of value by other flesh foods, then dairy products, with vegetables bringing up the rear. Indeed, fresh fruit has often been damned as harmful, or at best extravagant, by those who extolled meat's virtues – a prejudice that endured from the fifth until well into the nineteenth century. The

Washington Board of Health even went so far as to ban the sale or importation of practically all fresh fruit and also most vegetables during the cholera epidemic of 1832, since these were said to be 'highly prejudicial to health' (Griggs 1986: 2, 6, 18). To this day fresh fruit and vegetables play only a bit part in the diet of keen carnivores, with meat and animal products in the leading role. A rough hierarchy of prestige – and it can be only a rough representation, since our intricate vocabulary of foods is as subtly inscrutable as any other language – can be seen in Figure 5 (p. 112).

For vegetarians, and many who regard themselves as health conscious, the hierarchy is essentially reversed, so that items towards the top of the list are of greatest worth. True vegans aim totally to avoid all animal produce – including dairy products, wool, leather, even honey, and sometimes extending to such items as batteries or beer which are reputed to use animal products in manufacture – since their production is thought inevitably to harm sensitive animals. One common vegan argument against eating dairy items, for example, is that milk production not only involves rearing the cows themselves in unnatural and uncomfortable conditions, but also necessitates the slaughter of surplus male calves:

> 'I actually get more annoyed with so-called vegetarians than I do with meat eaters. They behave as if they're perfect just because they're too squeamish to eat meat, but they don't care about the horrible horrible ways in which cows are treated to get milk out of them. No one seems to care about that.'

Most non-meat eaters will eat eggs, cheese, or milk – foods which may be seen as the 'fruits' of the animal world, since they can be won without (necessarily) physically harming the donor. We would all consume such animal produce in a fabled land of milk and honey: when living in utopian harmony with abundant nature. For the more scrupulous, however, vegetarian cheese may be made with a substitute for the traditional curdling agent of rennet which is derived from the stomachs of slaughtered calves. Many who call themselves vegetarian will eat shellfish, or indeed larger seafood (although others are squeamish about shellfish, perhaps due to their being considered 'marine insects'). Some will even permit themselves poultry, with only red meat – the archetype of meat – entirely taboo.

Figure 5 A hierarchy of foods

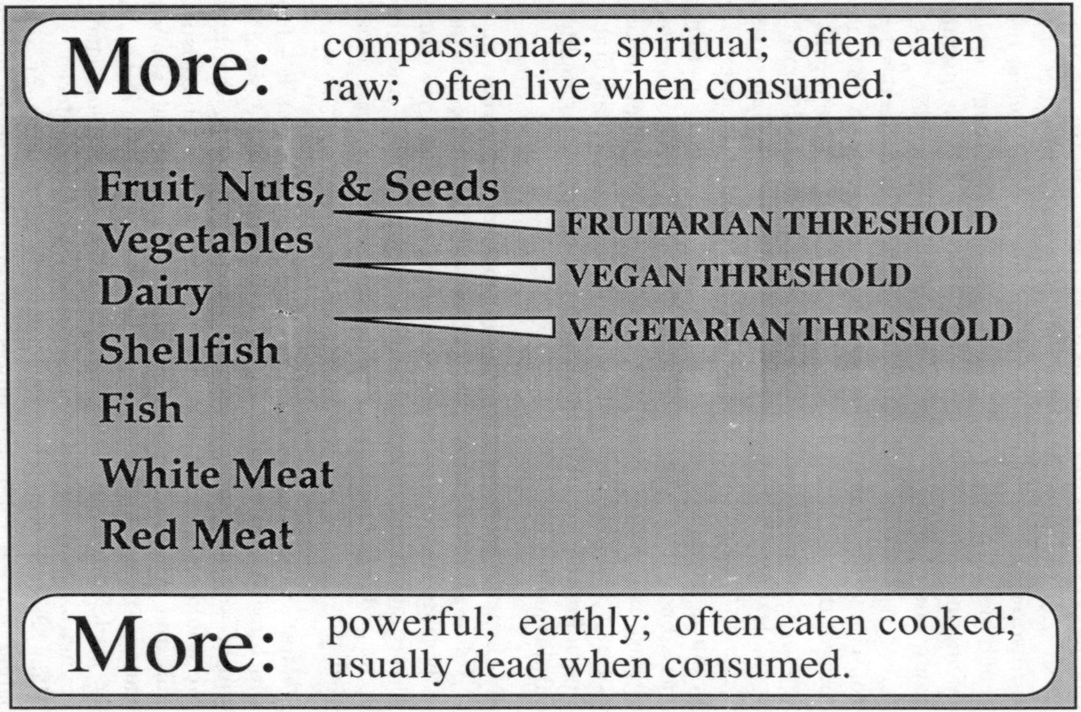

More: compassionate; spiritual; often eaten raw; often live when consumed.

Fruit, Nuts, & Seeds

FRUITARIAN THRESHOLD

Vegetables

VEGAN THRESHOLD

Dairy

VEGETARIAN THRESHOLD

Shellfish

Fish

White Meat

Red Meat

More: powerful; earthly; often eaten cooked; usually dead when consumed.

Amongst zealous vegetarians, however, in contrast to meat eaters, the most valued foods are typically nuts, seeds, and fruit: in other words the parts of the plant which can be eaten with *least* harm to the survival of the organism, as well as being lower on the food chain and so less ecologically costly in terms of land and resources. Similarly, such people will typically extol staple crops such as wholegrain rice as an ideal food of the highest nutritional and spiritual value, in contrast to its processed equivalent's reputation as a mere fill-you-up in the conventional scheme of thinking:

> Whole-grain or brown rice is referred to as 'soul food' by [Macrobiotic guru] Ohsawa who said it had the perfect balance of yin and yang. You are not expected to live on rice alone, but it will certainly be an important, central part of your diet. It is convenient for daily use, and it will help calm you when you are feeling edgy from dietary overindulgence or in times of stress or sickness.
>
> (Michell 1987: 55)

One pole of the list (Figure 5) is associated with life, the other with death. Non-meat eaters in particular will often explicitly contrast the vitality of fresh foods with the decay of animals' corpses. 'Dead' is a term similarly applied to over-processed industrial foods, including those sterilised by nuclear irradiation. Living foods, by contrast, are praised – particularly seeds when they have begun to sprout and the life forces been activated. Indeed, some fruitarians restrict themselves entirely to such items, and fresh fruit and raw vegetables are central to many healing diets in systems of complementary medicine.

Meat eaters, on the other hand, will understandably insist on the beast being dead before it is cooked or consumed. The threshold of acceptability for most modern westerners probably falls with 'live' dairy products such as yogurt, although cooking lobster by immersion into boiling water whilst still alive remains widely accepted and a minority will enjoy eating live shellfish – clams on the halfshell, for example. Even the most enthusiastic of western carnivores, however, would be unlikely to eat anything lower on Figure 5 – or even a part of it – while it was still alive, except within the safe bounds of humour:

> 'I suppose you've heard the one about the pig with three legs? This guy asks the farmer why his pig is wandering

around the courtyard on three legs: "Did it have an accident?", he asks.

'"Well, not exactly", says the farmer. "It was like this, you see . . . It was last year, and my little boy was playing down there by the river and he fell in. And, you see, no one heard his cries, except this pig here. So it leapt out of its pen, ran down to the river, dived in and fished the boy out, gave him the kiss of life, and then rushed off to get help".

'"So it broke a leg on the way, did it?", asks the man.

'"No, no, no. You see, then there was the time when my wife was taking the baby down into town, and she tripped, you see. She lost hold of the pram, and it went away off running down the hill towards the main road. The pig heard her let out an almighty cry, and came running to see what the trouble was. So it dashed down the hill, and caught the pram, it did, just as it was almost under the wheels of a juggernaut".

'"But the lorry hit its leg, did it?", suggests the man.

'"Oh no, it wasn't that at all. Then there was this other time too, when we were all in the fields, and a fire broke out in the kitchen. And this here pig smelled smoke, you see, and went and got buckets of water from the well and doused the flames, and then telephoned the fire brigade for help. Saved our farmhouse, that pig did!".

'"But how did it lose its leg?" asks the man yet again.

'"Well, you see", answers the farmer, "you've got to understand, when you've got a pig like that, you don't go eating it all at once".'

We have already noted that foods towards the base of Figure 5 are rarely consumed without having first been cooked or otherwise acculturated through processing such as smoking. Only in exceptional circumstances will we eat meat or fish raw, as a statement of either brute potency (steak tartare) or perhaps fashionable sophistication (sushi). Cooking ameliorates the stark animality of the flesh, by altering its colour, imposing a human hallmark since we are the only species to possess this skill, and confirming, beyond doubt, the death of the beast. Fruits, on the other hand, are most often eaten raw, as if not to destroy their life force.

Since cooking, particularly of meat, is a universal human trait, it is surely significant that even this degree of processing should be denounced as unhealthy by some. Cooking still denotes civilisation, but this can also come to be a negative rather than a positive indicator, again reversing the normal symbolism. Even at the beginning of the nineteenth century, Shelley was writing in his philosophical notes to *Queen Mab* that 'Prometheus (who represents the human race) effected some great change in the condition of his nature, and applied fire to culinary purposes; thus inventing an expedient for screening from his disgust the horrors of the shambles. From this moment his vitals were devoured by the vulture of disease' (Shelley 1813: 920–921). Certain modern approaches to healthy dietary practice, likewise, argue that raw vegetable foods are a better basis for life than cooked animal foods:

> No one would question that cooked foods have the ability to sustain life. What is questioned by doctors and scientists involved in research into raw diets is whether cooked foods are capable of regenerating and enhancing health. For, unless the genetic inheritance of a person is exceptionally good, a diet too high in cooked foods can lead to slow but progressive degeneration of cells and tissues, and encourage early aging and the development of degenerative diseases. Why? Some of the reasons no doubt depend on the fact that many essential nutrients are destroyed by cooking. Studies have shown that food processing and cooking – particularly at high temperatures – also bring about changes in the nature of food proteins, fats and fibre which not only render these food constituents less health-promoting to the body, but may even make them harmful.
>
> (Kenton and Kenton 1984: 35)

The medical sciences have weighed in with evidence to suggest that it is the cooking of meat, rather than the meat itself, which may sometimes be harmful. One such piece of research from Washington University, Missouri, for example, postulates the existence of 'culinary carcinogens', noting that the mutagen content of well browned hamburgers was fourteen times greater than lightly cooked ones (Eagle 1978: 12). Cancer, more than any other illness, is understood to be a 'disease of affluence', induced by the environmental pressures of modern industrial living. It is strangely fitting that eating cooked meat should increase its risk, since meat represents the power of civilisation, whilst cooking transforms nature's harvest into a product of human industry.

Meat now signifies barbarity as well as power, and this meaning permeates countless contexts. The major hamburger retailing chains, for example, have been tending to reduce the proportion of the colour red in their decoration and advertising. Meat, as already noted, is intrinsically linked with red blood – but the colour carries with it a series of associations, largely concerning power, violence, and danger. These are ideas which the fast food purveyors would be keen to curtail, since part of the burger's attraction is its sanitised supply. Instead, pastel shades present a gentler image than that of fiery, savage red. Sometimes even green is employed – the colour of chlorophyll – which stands increasingly for nature, for health, for freshness. It is the colour adopted by environmentalists who campaign for political and economic policies more attuned to the harmonies of nature. And, curiously, the complementary colour of green (the colour which you will 'see' if you stare at a green light and then close your eyes) is red: green's symbolic opposite, on the face of a traffic signal, and elsewhere.

The old orthodoxy, which saw power as an unqualified boon, has been falling out of favour. This has happened as concern has grown about the negative social and ecological implications of that ideology permitted to operate unchecked. Those who hold to the conventional assumption that human interests automatically override the interests of any other species are increasingly accused of arrogant hypocrisy and injustice. And concern is no longer restricted to a few well-meaning people who can be conveniently marginalised. Even the voices of such establishment figures as members of the British Royal Family are heard to call for new thinking. Ill-treatment of animals, or of the environment in general, may be seen not just as wrong, but as tyrannical or uncivilised, with comparisons drawn with other once accepted practices that are now beyond the pale. Marjorie Spiegel, for example, develops an extended comparison between human slavery and our use of 'lower' animals:

> As long as humans feel they are forced to defend their own rights and worth by placing someone beneath them, oppression will not end. This approach, at the very best, results only in an individual or group of people climbing up

> the ladder by pushing others down . . . Only through a rejection of oppression and institutionalized suffering *themselves* will we ever find a long-term freedom and justice *for all* . . . A line was arbitrarily drawn between white people and black people, a division which has since been rejected. But what of the line which has been drawn between human and non-human animals? . . . The more we learn about the earth's environment, its ecosystems and the creatures who live here, the more we see the absurdity in the concept of ranking species against one another . . . Each species has attributes which others lack, and it is only an anthropocentric world view which makes qualities possessed by humans to be those by which all other species are measured.
>
> (Spiegel 1988: 16, 17, 19)

Within this frame of reference consideration for the other inhabitants of our planet is regarded as a mark of cultured development and transgressions against such standards as less acceptable behaviour. The novelist Richard Adams, continuing this theme, paraphrases Jeremy Bentham's well-worn dictum:

> The luxury fur industry represents by far the worst abuse of sentient, warm-blooded mammals at present condoned by law in the Euro-American world: first, on account of its large scale; second, on account of the gross cruelty involved; and third, on account of the unnecessary nature of the end product . . . The time is not far off when the fur trade will be universally seen, like slavery and child labour, as a barbarous anachronism . . . What is it worth to our collective self-respect to abolish this stigma? £50 million? If you feel it's not worth any economic sacrifice at all, then perhaps your self-respect needs updating. The question is not can animals reason or communicate. It is: can they suffer?
>
> (Adams 1989: 9)

The new ideology, in contrast to tradition, regards unrestrained domination of other creatures as a sign not of civilised elevation but of regrettable backwardness. Activities such as hunting, or the ostentatious display of fur garments, whereby we have demonstrated our mastery for so long, are disparaged accordingly. If this orientation were restricted to a few 'animal rights activists' and ethical campaigners then its significance might be limited, but that is by no means the case. In 1990 three-quarters of British women are said that they would not wear animal fur, and British supermarkets have begun to announce that they will now be labelling 'cruelty-free' products throughout their stores. Elements of such ideas have diffused throughout our population. The implications for meat production are considerable should this recent reappraisal of values continue.

Part III
MIXED MEATAPHORS

8
THE RELUCTANT CANNIBAL

Going around saying 'Don't eat people'
Is the way to make people hate you
Always have eaten people
Always will eat people
You can't change human nature!

('The Reluctant Cannibal' from *At the Drop of a Hat*
William Flanders and Donald Swann)

Human flesh, nutritious though it may be, is not normally on our menus. This may seem somewhat obvious, but our inclination not to eat each other is a vital piece in the jigsaw of beliefs which inform our views on that which we can and do eat.

We frequently hear, of course, of barbarous societies in other times and other places who have practiced cannibalism. We are reared on endless stories of such savages from an apparently comprehensive literature by authorities such as explorers, missionaries, anthropologists, colonial administrators, and archaeologists. It comes then as a surprise, if not an affront to common sense, to hear it suggested that there may in fact be no 'adequate documentation of cannibalism as a custom in any form for any society' (Arens 1979: 21).

In *The Man-eating Myth*, Arens reviews the documentary evidence for cannibalism and reports himself unable to unearth a single credible eyewitness account of the act occurring as a customary event, anywhere, ever. He excludes aberrant individual behaviour or cases under survival conditions, such as sieges or the aftermath of the aircraft crash in the Andes mountains when some survivors resorted to consuming those who had perished (Read 1974). But each time ritual cannibalism is said to have occurred, Arens claims, investigation reveals that the supposed anthropophagy is based on either hearsay or plagiarism, or is only reputed to have occurred in the recent or distant past, or amongst other neighbouring or distant societies, or else is based on manifestly unreliable testimony.

The phenomenon, he argues, is so well known to exist that its occurrence is repeatedly assumed *prior* to consideration of the circumstances of a particular report, instead of its being approached with due scepticism. Arens illustrates his argument with celebrated instances of cannibalism, ranging from Hans Staden's sixteenth-century account of the practice amongst the Tupinamba in South America, which he suggests is not only untenable but may itself be the source of many succeeding reports phrased in suspiciously similar terms; to Aztec man-eating, which was likewise, he says, reported by no eyewitness; to more recent reports of cannibalism from Africa and New Guinea (including Gajdusek's Nobel prize-winning work on transmission of the kuru disease amongst the Fore people), to archaeological evidence. Not one example survives his scrutiny.

Although it is logically impossible to disprove the existence of any social practice, since a single properly documented instance could overturn any such argument, Arens's analysis and conclusions should convince most readers that the phenomenon of cannibalism is, at least, much more rare than is generally believed. The significance of his work, however, lies not only in his incisive perspective on the reports themselves, but in his consideration of the phenomenon as a whole and his willingness to turn the spotlight around to observe the observer:

> The most certain thing to be said is that all cultures, subcultures, religions, sects, secret societies and every other possible human association have been labelled anthropophagic by someone. In this light, the contemporary, though neglected, anthropological problem emerges more clearly. The idea of 'others' as cannibals, rather than the act, is the universal phenomenon. The significant question is not why people eat human flesh, but why one group invariably assumes that others do. Accounting for a single aspect of an overall system of thought, rather than an observable custom, becomes the issue.
>
> (Arens 1979: 139)

He resolves this problem by looking with fresh eyes at the common characteristics of our own views on the matter, where he notes three points in particular: firstly, the 'basic notion that customary cannibalism not only still exists but was once much more pervasive. Second, the subject matter is mystified by resorting to a specialized vocabulary . . . Third, the objects of all the intellectual energy are the primitives'. . .

> Much to our satisfaction, the discussion of cannibalism as a custom is normally restricted to faraway lands just prior to or during their 'pacification' by the various agents of western civilization . . . 'they,' in the form of distant cannibals, are reflections of us as we once were.
>
> (Arens 1979: 18–19)

Here is the crux of the matter. Defining other people as cannibals, Arens argues, is generally an instrumental act whereby the alleged perpetrators are placed outside the realm of civilised culture. Cannibalism is often thus an unwarranted, but widely useful, instance of collective prejudice:

> [an] aspect of cultural-boundary construction and maintenance. This intellectual process is part of the attempt by every society to create a conceptual order based on differences in a universe of often-competing neighbouring communities. In other words, one group can appreciate its own existence more meaningfully by conjuring up others as categorical opposites.
>
> (Arens 1979: 145)

And in this respect, he says, we are in no major respect different from the 'primitive' peoples upon whom we pass implicit judgement. 'However, as befits a complex society, we have the services of a distinct scholarly discipline to systematize the simple notions which must serve among primitive peoples' (1979: 169).

Everyone, it seems, has believed others to be cannibalistic. A reflection of the medieval English mind, for example, is represented in the Mappa Mundi (*c.* 1290 AD) and, sure enough, somewhere west of the Black Sea we find 'Cannibals – or "Essendones" who reputedly ate their parents' flesh' (Jancey 1987: 14). Shakespeare, too, used the device in conjunction with physical deformity to impute sub-human status to peoples encountered abroad; his hero recounts memories:

> . . . of the Cannibals that do each other eat,
> The Anthropophagi, and men whose heads
> Do grow beneath their shoulders
>
> (*Othello* I.iii)

And accusations of cannibalism are still regularly employed to designate others as uncivilised. A modern example of this symbol used to its full effect provides the climax to Peter Greenaway's 1989 film *The Cook, the Thief, his Wife, and her Lover,* when the peerlessly brutal and boorish thief (Michael Gambon) is forced to dine on the corpse of one of his victims, cooked and exquisitely garnished, to fulfil his prior taunts. This final depravity, which nauseates even him, serves to substantiate his ultimate debasement, isolating him beyond the bounds of humanity, and so to endorse his ignoble death.

The racism that is often implicit in accusations of cannibalism is evident in a tabloid newspaper's front-page feature in the midst of political controversy about British national representation at the funeral of Emperor Hirohito of Japan, still believed by many to have been a war criminal. 'JAP WHO ATE OUR TROOPS FOR DINNER', screams the headline, reporting US war archives reportedly documenting the eating of a dead British airman at a 'Japanese general's jungle cocktail party', one of 'no fewer than EIGHT files giving horrifying accounts of Japs eating human flesh' (Trueman 1989: 1, 5). The implied sub-humanity of the entire culture effectively incriminated is indeed spelled out, in case the reader be left in doubt:

> News of the appalling atrocity revolted British Far East veterans last night.
>
> But a spokesman for the Burma Star Association said: 'The news does not surprise us one bit.
>
> 'We always knew they were animals – now we have the proof that they were cannibals'.

In the light of Arens's arguments, however, it is interesting to note that the evidence – reportedly given by a Japanese officer on trial for war crimes to a US military tribunal after the war – states that the cases occurred 'when the Japanese Army ran short of food in 1945', and may therefore have been (if it indeed ever took place) under what might be termed survival conditions. But more significantly still, it is eventually clear that the report is again not based upon a full eyewitness account:

> . . . the general mentioned to me that there had been an execution that day and we should send for some meat.
>
> 'I took that to mean the body of the Allied pilot . . .
>
> 'Everyone had a taste . . . and although I wasn't specifically told I always understood we had eaten pieces of the allied flier.'

Alternatively, Arens notes, an individual or group can derive kudos from describing *themselves* as cannibals, the claimed transgression of such basic standards suggesting their ultimate power or lawlessness. Thus, in another feature film, *Crocodile Dundee II,* two Australian aborigines enact a charade to unnerve the villainous American prisoners they are guarding on behalf of the hero. One asks his companion whether they are allowed to eat their captives and is told that no, they are just to be held . . . as a wink is passed to reassure the heroine nearby. The psychological gambit succeeds in demoralising their charges, one of whom crosses himself in despair. Similarly, a report in a British newspaper, even if believable, does not in fact describe an act of cannibalism – it only relates a claim, that might be no more than a gesture of bravado, by one marginal individual with a suitably demonic name:

> Swaziland is to deport a Moroccan cannibal who has been living in the kingdom as a refugee. Hitler Sharin, who has just completed a six-month jail sentence for illegal possession of weapons, demanded the bodies of road accident victims for his meals. The Swazi authorities refused to meet the demand.
>
> (*Edinburgh Evening News* 1990: 5)

Normally, the contrast is drawn between opposing cultural groups, the cannibalism label being used to denote a different level of humanity. Occasionally, similarly, it crops up ostensibly describing our uncivilised past. An occupational psychologist, for example, uses the motif to illustrate the importance of first impressions in job interviews:

> But what are those crucial, early impressions based on? 'Probably on unconscious body language of the kind that

> was useful when we lived in the jungle – meeting a stranger we hadn't long to make up our mind whether to eat them or share our food with them. Most human beings make very fast decisions about people'.
>
> (Flowers 1989: 19)

The argument rumbles on. Peggy Reeves Sanday thinks that Arens overstates his case in 'arguing that cannibalism has never existed' (although he does not, in fact, say this, but merely that the evidence for its existence as an accepted practice is inadequate). She refutes his scepticism with further hearsay (Sanday 1986: 9–10). She examines a 'representative sample' of 156 societies and detects cannibalism in 34 per cent of those where there was sufficient evidence to make a judgement; few of her examples even purport, however, to be based on eyewitness evidence, and many are mere snippets of conversation.

The *London Review of Books* hosted a dispute on the matter when Chinua Achebe described Joseph Conrad as 'a thoroughgoing racist' for characterising inhabitants of the Congo in the nineteenth-century as cannibals in *Heart of Darkness* (Achebe 1988: 8). His reviewer, Craig Raine, defends Conrad's belief in cannibalism, suggesting that Achebe should 'accept the uncomfortable fact of its existence' (1989: 17); he cites several sources to justify his certainty including an illuminating passage from the trader and explorer H.M. Stanley:

> For the most part it [the Congo] is peopled by ferocious savages, devoted to abominable cannibalism and wanton murder of inoffensive people, but along the great river towards Livingstone Falls there dwell numerous amiable tribes who would gladly embrace the arrival of the European merchant.
>
> (quoted in Raine 1989: 17)

The 'fact' that Stanley's cannibals were defined as those who refused to enter into so-called free-trade with him casts doubt on his objectivity. Similarly, as Patrick Parrinder suggests in a reply to Raine in the *London Review,* the failure of explorers such as Stanley, Grenfell, and Bentley to encounter cannibalism at first hand is significant. It recedes like a mirage wherever approached, although always well known to occur nearby and to have been rife when the missionaries first arrived. Why? Parrinder suggests that it was the self-sacrifice of those European missionaries who died there, in numbers that required justification, which was amply supplied by stories of cannibals who abandoned their ghastly practices under Christian teaching. But, he counsels, 'if cannibalism was really a socially approved . . . practice, we must ask how it came to be given up so easily' (Parrinder 1989: 4).

The principal difference that separates Arens's analysis from most is in the quality of evidence tolerated; Sanday, for example, apologetically justifies her 'uneven' data as the 'best available' (1986: 10). The manifest unreliability of many reports does indeed present a dilemma to those who wish to unravel the practice, since if dubious cases are excluded precious few remain. To begin with, accounts of cannibalism by missionaries, commercial speculators, or other parties with a vested interest in portraying the accused as in need of 'civilisation' should certainly be treated with scepticism. Reports based on hearsay, which means the vast majority of examples, should likewise be viewed with caution since such 'Chinese whispers' can easily distort the truth. One should also hesitate before believing claims of past cannibalism that could be mere posturing. And, even then:

> if cannibalism in the 19th-century Congo had been as prevalent as murder and rape in contemporary Britain, it would still be considered racist by today's standards to refer habitually to all Congolese, or to all Congolese of a particular tribe, as cannibals – unless, that is, reputable evidence of a general social approval of the practice existed.
> (Parrinder 1989: 4)

In disputing the prevalence of normal cannibalism, Arens directly criticises the integrity of generations of anthropologists who, he suggests, have been marvellously naïve in accepting its mythology as documented fact. He hints, moreover, that this willing acceptance reflects general racism. Cannibalism has also been a glamorous topic for anthropologists to write about. Predictably, this damning indictment has not been widely welcomed within the discipline. In a review article in *Man,* for example, Peter Rivière calls it a bad book and 'also a dangerous book' (1980: 205) and, in the preface to a volume on the ethnography of cannibalism, Tuzin and Brown simply say that he is wrong (1983: 3). They nonetheless concede the plausibility of Arens's 'suggestion that the common *attribution* of cannibalism is a rhetorical device used ideologically by one group to assert its moral superiority over another' (1983: 3), even if they cannot accept that anthropologists could possibly be one such group.

Perhaps the most eloquent testimony to Arens's contribution, however, is that most discussion of cannibalism is now in terms of the symbolic or ideological dimensions of the alleged practice, rather than seeking to validate its authenticity. Thus Poole concludes that for Bimin-Kuskusmin, 'the idea of cannibalism implicates a complex amalgam of practice and belief, history and myth, and matter-of-fact assertion or elaborate metaphor' (1983: 31); Sahlins characteristically finds that 'the historical practice of cannibalism can alternately serve as the concrete referent of a mythical theory or its behavioral metaphor' (1983: 91); and Sanday acknowledges that it 'is never just about eating but is primarily a medium for . . . messages having to do with the maintenance, regeneration, and, in some cases, the foundation of the cultural order' (1986: 3).

Whilst the factual basis of many reports might be questioned, there is no doubt that cannibalism is still regularly reported, and that is itself meaningful. The importance of Arens's revelation is considerable. If he is correct, normal aversion to eating the flesh of our own species is – if not universal – at least much more general that we habitually believe. This should not, in fact, be surprising, since few animals consume the flesh of their own species, even after death, other than in exceptional circumstances. If the taboo is indeed prevalent, the very potency of any report of cannibalism reflects the depth of feeling with which we maintain the proscription against this practice. Indeed, perhaps what is most significant about the aircraft crash in the Andes (Read 1974) is not that a few survivors were willing to resort to cannibalism but that several would not do so, so strong was their aversion, and perished as a result. If Arens is right to doubt that cannibalism has ever been common, one function of the taboo has been to demarcate the sanctity of human society. Our culture is distinguished as civilised, as a higher form of life that cannot be preyed upon. Our humanity itself is quite literally at stake.

Whether or not anthropophagy is indeed universally proscribed, it is clearly not *our* normal practice. As one aspect of our classification of potential foodstuffs, this rather conspicuous fact is of central significance. The taboo involves far more than just whether or not we would enjoy consuming human flesh, not least since defining what is *not* edible carries logical implications as to that which *is* edible. At the broadest scale of resolution, the assumption that we do not eat other humans suggests, by simple opposition, that anything non-human is potential food – animal, vegetable, or mineral – unless proscribed for other reasons.

That, however, is an untenably broad level of analysis. In practice, classification is very much more complex. In the next two chapters we will see how these notions, of inedible, civilised human beings opposed to the edible, primitive wilds of nature, extend to permeate our thinking in some unexpected ways. Meanwhile, it is worth noting how cannibalism can be invoked to dispute conventional wisdom on the eating of meat.

Since belief in the edibility of non-human animals depends upon a clear conceptual division between them and ourselves, it is perhaps not surprising that attempts to contest the justice of meat eating commonly refer in some way to cannibalism. Such arguments typically confront the boundary between them and us as an arbitrary or inappropriate distinction, pointing up the similarities between species, rather than the differences, to make their consumption seem less acceptable by bringing them conceptually into the fold of humanity. George Bernard Shaw, for example, described meat eating as 'cannibalism with its heroic dish omitted' (Cox and Crockett 1979: 54). Whenever, likewise, 'murder' is applied to the killing of a non-human animal, there is an imputation of quasi-human status to the animal concerned. The Smiths pop group, for example, recently entitled one of their top-selling record albums *Meat is Murder* – a phrase that can now also be seen on innumerable badges.

A fine example of this brand of argument, worthy of extended treatment, is Elisee Reclus's classic polemic 'On Vegetarianism', which first appeared in the *Humane Review* in January 1901. Throughout the piece Reclus makes reference, in one form after another, to the implication of cannibalism. He starts by recounting his first encounter with butchery as a small boy, comparing his own frame to that of a carcase . . .

> I seem to have heard that I fainted, and that the kind-hearted butcher carried me into his own house; I did not weigh more than one of those lambs he slaughtered every morning.
>
> (Reclus 1901: 2–3)

. . . which he rapidly counterpoints by likening the sound of a dying pig to a child:

> She cried without ceasing, now and then uttering groans and sounds of despair almost human; it seemed like listening to a child.
>
> (Reclus 1901: 3)

He proceeds to make explicit the connection he perceives between the eating of meat and the abrogation of civilised status, invoking the barbaric conduct of our soldiers at war in China . . .

> But is there not some direct relation of cause and effect between the food of these executioners, who call themselves 'agents of civilisation,' and their ferocious deeds? They, too, are in the habit of praising the bleeding flesh as a generator of health, strength and intelligence. They, too, enter without repugnance the slaughter house, where the pavement is red and slippery, and where one breathes the sickly sweet odour of blood. Is there then so much difference between the dead body of a bullock and that of a man? The dissevered limbs, the entrails mingling one with the other, are very much alike: the slaughter of the first makes easy the murder of the second . . .
>
> (Reclus 1901: 6)

. . . until finally, grasping the nettle, he declares that other animals should be accorded like consideration:

> But however this may be, we say simply that, for the great majority of vegetarians . . . the important point is the recognition of the bond of affection and goodwill that links man to the so-called lower animals, and the extension to these brothers of the sentiment which has already put a stop to cannibalism among men. The reasons which might be pleaded by anthropophagists against the disuse of human flesh in their customary diet would be as well-founded as those urged by ordinary flesh-eaters today . . . The horse and the cow, the rabbit and the cat, the deer and the hare, the pheasant and the lark, please us better as friends than as meat. We wish to preserve them as respected fellow-workers, or simply as companions in the joy of life and friendship.
>
> (Reclus 1901: 8)

The entire argument is effectively emotive since it strikes at the heart of some of our most fundamental assumptions: that humans are not to be eaten, that other animals potentially are, and that there is a clear dividing line between the two categories. By challenging us, through various devices, to consider the possibility that the distinction is not as clear as we habitually assume, Reclus attempts to persuade us to extend some of the same consideration that we like to believe we have for other humans, to other animals. Meat, he says, is murder.

9

PETS AND OTHER GREY ANIMALS

The rabbit has an unusual place in our categorisation of animals, since it is seen by some as a pet, and by others (or even by the same people) as edible. However, we do not normally eat pets:

> 'I remember hearing about an advertisement once in the States, when someone had put a small-ad in the paper for "Rabbits for sale: as pets, or for the freezer", and there was an amazing row about it . . . people couldn't handle the idea of both things in the same advert.'

An advertisement for either purpose would pass largely unnoticed, but to suggest both possibilities in tandem is too direct a reminder of an anomaly in our classificatory system, and stimulates howls of protest from an outraged public. We do not normally eat pets. In Britain we do not eat dogs or cats, nor canaries, nor goldfish. On an infamous episode of television's *Candid Camera*, for example, the presenter shocked members of the public by pretending to snatch and eat fish-shaped pieces of carrot from of a fishtank. The very idea of eating pets can be enough to put us off meat entirely:

> 'I once watched this film about . . . gangsters or something . . . and they made this man eat his pet poodles, because he really liked his pet poodles, so the gangsters mixed them up and made him eat them – pretended it was chicken or something – and then after he'd eaten them they brought out the silver plate and lifted up the lid and there was these poodles' heads underneath. And I could never eat minced chicken after that, but apart from that I always thought that human flesh would taste a bit like poodles. That probably

didn't help – mashed up poodles on the television – I felt a bit sick after that film. That definitely put me off chicken.'

Why should someone (who presumably has never sampled it) suspect that poodle meat might taste similar to human flesh? Pet dogs are not human. But they are the next best thing. We exclude pets from our food resources due to their social proximity to ourselves as humans. As Simoons suggests,

> familiarity with animals, particularly in functional relationships and as pets, led to the rejection of entire species of domestic animal. Avoidance of dogflesh in the Western world may have come about because the dog was the friend of the family and eating it seemed an act akin to cannibalism
>
> (Simoons 1967: 114).

And just as some campaigners evoke cannibalism as a device to dispute the legitimacy of eating other animals, so our attitude towards pets can be used to challenge conventional assumptions:

> *Q. 'Do you ever get 'attacked' by people for being vegetarian?'*
>
> 'Yes, I've had lots of arguments of that sort, but not particularly passionate ones usually; it tends to be more academic. I usually come out with something like "well, would you eat your pet dog?"'

By caring for pets, tending them, giving them proper names, we endow them with semi-human status. The foods we allot them are largely modelled on human tastes: no manufacturer markets the mouse or bluebottle flavour cat food for which, given free choice, eight out of ten cats might express a preference (nor cat-flavoured dog food!) We allow pets into our houses, and sometimes even into our beds; we talk to them; we give them special affection, special medical care, special exercise; we fret when they are unwell and weep when they die; we may even bury them alongside us.

We treat pets more like individual subjects than the abstract objects as which we officially regard edible animals – although some species we treat more favourably than others, allotting them a closer relationship to ourselves. Few in Britain would enjoy consuming horse, although that is probably less unthinkable than roast of dog. We tolerate those who do eat horse, such as the

French and Belgians, with amused disdain, but are troubled to learn that they import *our* horses for consumption. 'I could eat a horse' effectively expresses extreme hunger not only due to the large size of the animal, but also because it implies a willingness to transgress normal standards of edibility on account of desperate need. Even then, however, we are unlikely to suggest that a slice of Puppydog Pie would make a tasty starter:

> as domestic cohabitants, dogs are closer to men than are horses, and their consumption is more unthinkable: they are 'one of the family.' Traditionally horses stand in a more menial, working relationship to people; if dogs are as kinsmen, horses are as servants and nonkin. Hence the consumption of horses is at least conceivable, if not general, whereas the notion of eating dogs understandably evokes some of the revulsion of the incest tabu.
>
> (Sahlins 1976: 175)

As honorary humans, pets cannot be consumed. As Marshall Sahlins puts it, 'To adopt the conventional incantations of structuralism, "everything happens as if" the food system is inflected throughout by a principle of metonymy, such that taken as a whole it composes a sustained metaphor on cannibalism' (1976: 174). The metaphor is sustained because it affirms the sanctity of the boundary between us and them, between human and non-human, between subject and object, between civilisation and its resources.

The ideological sensitivity of the distinction is shown by the occasional public outcry at scientific activities which threaten the divide. The spectre of men-monsters of strength but low intelligence – a new slave race, and always a staple of science fiction – are now a real possibility with advances in genetic engineering. But we are distinctly uncomfortable with the concept of creatures which cannot be allocated neatly to one category or the other. We know that humans have rights and must be treated with appropriate respect since slavery, for example, has been officially outlawed. We also know that lower animals may be put into our service in whatever way we see fit. But we do not know precisely how we should perceive a being which straddles the classifications. We might prefer, therefore, that such a creature were not invented.

In similar terms, we know that humans are not fit food. We

know that non-human animals, unless otherwise proscribed, are. So any beast that falls between human and non-human, by coming close in some way, tends to be deemed inedible, lest it make us doubt our certainties. But viewed through the looking glass, even a formal introduction may breach the chasm that separates us:

> At last the Red Queen began. 'You've missed the soup and fish,' she said. 'Put on the joint!' And the waiters set a leg of mutton before Alice, who looked at it rather anxiously, as she had never had to carve a joint before.
>
> 'You look a little shy: let me introduce you to that leg of mutton,' said the Red Queen. 'Alice – Mutton: Mutton – Alice.' The leg of mutton got up in the dish and made a little bow to Alice! and Alice returned the bow, not knowing whether to be frightened or amused.
>
> 'May I give you a slice?' she said, taking up the knife and fork, and looking from one Queen to the other.
>
> 'Certainly not,' the Red Queen said very decidedly: 'it isn't etiquette to cut anyone you've been introduced to. Remove the joint!'.
>
> (Carroll 1872: 240)

Modern western society is by no means unique in avoiding the flesh of over-familiar animals. Jewish custom treats working animals more like pets, and similar notions are found among the Bari of Sudan, where the owner of an elderly ox will be grief-stricken by its eventual killing, as well as in modern Greece, China, and Korea. Abstention under these circumstances, 'though it may derive in part from a desire to preserve useful animals . . . is also motivated by affection for a companion or friend' (Niven 1967: 10–11; Simoons 1967: 113). Simoons is correct, but fondness alone is insufficient explanation. Its effect comes because emotional attachment can undermine our clear categorisation of Human above Animal.

Another class of animals that most westerners regard as inedible is the primates, including monkeys, gorillas, chimpanzees, and even smaller primate species. The reason again relates to their closeness to humans, although in this case the proximity is morphological or physiological more than social. Looking at primates we recognise something of ourselves. They look like us. Otherwise, however, our reasons for not eating primates are precisely the same as for not eating pets: the infringement of the cannibalism taboo. The threat is to the world view which places us above and in control of the rest of nature, and which permits us to exploit as we will with moral impunity. The naturalist William Bingley observes, of monkeys, that there is:

> something extremely disgusting in the idea of eating what appears, when skinned and dressed, so like a child. The skull, the paws, and indeed every part of them remind us, much too strongly, of the idea of devouring a fellow-creature.
>
> (Bingley 1824: i. 86)

Once again, we in the modern western world are not alone in this trait. Tambiah, for example, relates the in-some-ways-similar attitude of Thai villagers towards the monkey which, they say, is because the monkey is 'descended from man'. They tell a story of when a poor woman's children went into the forest in search of food, and in time grew hair on their bodies:

> Monkeys are thus in a sense lost and degenerate human beings; their affinity to humans make them improper food. Yet it is whispered in the village that some people do eat them. Their animal and semihuman status is a bar to open cannibalism.
>
> (Tambiah 1985: 191)

This mention of cannibalism bears intriguing parallels with our tales of people eating pets, such as in 'modern myths' – stories passed from person to person, purportedly as second or third-hand experience. These share with 'proper' cannibalism the feature that those accused of transgressing decent norms are usually marginal to mainstream society. One example is the married couple visiting a restaurant, typically either abroad or run by members of a foreign ethnic group, who use sign language to ask that their pet be fed too. The dog is led off to the kitchen, and only too late do they discover that, due to misunderstanding, they have been given it to eat. A variation on this theme was related to me by one businessman as a memory of his home town (which enquiries into local sources failed to corroborate):

> 'I think I'd draw the line at eating cats and dogs. But that doesn't stop Indian restaurants from serving dogs . . .'

Q. 'Has that ever been proved, or is it another of these apocryphal tales?'

'Oh yes. There was one quite famous one of an Indian restaurant in my home town, where they were done for serving dog. I mean . . . there were certain benefits, like the number of strays there was kept to a minimum for a while.'

Revulsion at other societies' tastes, contrasted with the supposed rationality of our own, is commonly used to imply their lack of civilisation. Although many societies share the characteristic of avoiding species which are close to humans, which species are so privileged is by no means consistent from culture to culture. Many Africans and Amazonian Indians willingly eat monkeys, for example, and when preparing to host the 1988 Olympic games, the South Korean authorities took steps to prohibit local restaurants from offering the locally esteemed dish of cooked dog on their menus, for fear of displeasing the visiting tourists and media. Such measures failed, however, to appease western animal welfare campaigners who advertised in British newspapers with a photograph of a dog hanging by its neck:

> Watching TV coverage of the South Korean Olympics this Summer you'll see all the traditional scenes . . .
>
> You WON'T see the evil, uncivilised side of life in the host country. Outside Seoul's Olympic Park cats and dogs, just like our pets, will be brutally killed as *LUXURY* food for those who believe such dishes give them the strength and stamina of the animals.
>
> Kittens, cats and dogs will suffer appalling cruelty as they are slowly hanged . . . strangled . . . clubbed . . . or tossed alive into boiling water. Terror stricken animals, it's claimed, taste better.
>
> For five years IFAW has patiently pleaded with the Korean Government to end the barbaric practices.
>
> (IFAW 1988)

We are outraged at the uncivilised barbarism of those who eat the animals that we revere, yet we contentedly relish the flesh of a cow that to the orthodox Hindu is sacred. We are appalled that foreigners can treat dogs and cats so cruelly in the belief that this will improve their taste, yet our own shops sell prestigious veal whose distinctively tender white meat comes from immobilising a calf for life in a 5ft by 2ft crate so that it cannot suckle or groom, nor even lick the floor or its walls, denying it hay or fodder of any kind to supplement its all-milk crash diet; by the time it is ready for the journey to slaughter at the age of 12 weeks it may well be too weak to stand due to lack of exercise (Jackman 1989a: 47). We are nothing if not inconsistent.

Carnivores are also, apparently, not to our taste. In fact this avoidance is doubly curious since, according to a certain sort of economic logic, the flesh of carnivores ought to be the most highly esteemed of all, since it is the most difficult and dangerous to obtain, since such beasts are considerably rarer than the lower animals upon which they subsist, and since they are costly to rear. Until recently few objected to the hunting and killing of carnivorous species. Those who desired to display their grandeur have hung lions' and tigers' heads on their walls, or splayed the beasts' skins on floors, since slaying such prey traditionally bestowed great prestige. However, just as our society might sanction the killing of other humans under certain circumstances but never their consumption, whilst we may exert our authority in other ways we will not normally eat carnivores:

> *Q. 'How about eating . . . dog? Roast of dog?'*
>
> 'It wouldn't appeal to me, I must admit. I probably would, if there was nothing else about, if I was really really hungry . . . I think I would rather eat dog than grubs, or any sort of bug.'
>
> *Q. 'Why not dog?'*
>
> 'I don't know. I think it's that we're not used to eating any sort of carnivore meat, are we?'
>
> *Q. 'Why not?'*
>
> 'Because around here Man has done away with them . . . But would we ever have eaten foxes or wolves if they were killed, even going back in time? I doubt it very much. They would have used their fur, but not eaten them. I can see why we don't farm them certainly. It's basically just inefficient. Why feed them meat to produce meat? It just doesn't work, does it? They're only going to convert half of what they eat.'

The economic logic of inefficiency does not dissuade us from expending roughly ten times the food value on rearing domestic livestock than we receive in return from the meat, so it is difficult to see why this argument alone should preclude the eating of carnivores – particularly if we are shooting them anyway. Rather, there is simply something that seems not quite right about eating another predator.

A curious example of this syndrome is the controversy that developed in Britain in late 1988 about the level of salmonella contamination in eggs and chickens supplied to the public. In the course of media investigation it was revealed that a principle cause of the epidemic was that intensive producers had for some time been feeding the carcases of dead chickens, inadequately sterilised, back to other chickens as a protein supplement, so that infection was constantly circulated within and between flocks. This disclosure outraged members of the public who had been unaware of such practices. Concern seemed partly to stem from an uneasy feeling that by recirculating the ground-up remains of dead birds we were causing the chickens to be unnaturally cannibalistic, but also partly because this meant that they had now become carnivores. Even then, the poultry's new categorical status would surely have escaped widespread notice had they not been destined for our dinner tables.

The same pattern of concern, for similar reasons, repeated in 1989–90 over BSE in British cattle. It emerged that the source of the epidemic was likely to be their being fed with the remains of sheep to increase productivity; some of the sheep had presumably suffered from scrapie – a 'spongiform' disease endemic for centuries. Whilst scientific research focused upon the possibility of the agent transfering, in turn, to humans, an indignant public deplored the principle of feeding herbivores with animal remains at all. Cows do not naturally eat meat, and their being forced to do so by our profit-oriented agribusiness seemed to many to have brought its own punishment. But adding to the unease was an awareness that the cattle were now carnivores – and humans do not eat carnivores:

> *Q. 'You say you enjoy eating almost any meat, but . . . how do you think you would feel about eating, say, lion flesh, or wolf?'*
>
> 'Oh, I really don't know. Let me think about it . . . Er, no, I suppose if it came to the point and I was starving I'd

probably eat almost anything, but I think that's something I'd really prefer not to. I don't know why. I've never really thought about it like that. I just can't imagine that I'd like the taste. It would be far too strong, and probably pretty tough . . .'

Q. 'If you were actually offered some on a plate, would you try it?'

'Ha! To be honest I think I'd prefer not to. I'd have to be pretty damn hungry. I see what you mean . . . There's no reason not to try something, but I still think I'd prefer not to . . . It's not that I think it would actually be bad for me in any way, logically. It's just the thought of it which makes me think I'd feel sort of sick about it. I don't know actually – maybe there's something to do with their having eaten other animals . . . and so there's more chance of them having picked up something poisonous along the way or something? But no, I don't really think it's that. It's just the way I feel.'

The flesh of carnivores is commonly described, such as by the informant above, as tasting too strong to be eaten: a term that contains a clue to the source of the proscription. This, Julia Twigg says, 'is the familiar anthropological concept whereby that which is most highly prized, most sacred, can, by virtue of its power, be the most defiling'. She suggests that carnivores are 'like a double dose, too much of a good thing' (Twigg 1983: 22, 25). Certainly, it is not merely that those animals might be dangerous to catch – that might apply to some, but not to all, meat-eating species. Rather, carnivores are strong by being in an evident position of power over other animals that they are able to hunt and to eat. As such their place in the world is analogous to our own, since we regard ourselves as carnivores. The danger they present is to our minds as well as to our bodies:

> The fact that only those animals who somehow invert their own natural order, such as the renegade lion or tiger and certain species such as the solitary nocturnal leopard or hyena, sometimes prey on humans for food strengthens the symbolic association between cannibalism and antisocial behavior. Other species which in some way subvert the human interpretation of the natural order of things, such as the alligator, a reptile which inhabits the arena of fish, and

> the baboon, who physically parodies man and invades his domain for food, become other potential markers of evil. These are also the very species which human beings often exclude from their diet whenever possible because of their unsavory nature.
>
> (Arens 1979: 140–141)

In other words, once again, carnivores are close to us – not this time socially, nor morphologically, but functionally, and so to eat them would be similarly akin to cannibalism. (Of course some animals fall into more than one category, such as the especially privileged cats and dogs who are close to us both socially and due to their carnivorous habits.) We eat only animals which are 'natural victims', and carnivores cannot be consumed since their relationship to others is similar to our own. We give ourselves the right to kill our prey, and we accord carnivores the right to kill theirs. We respect them since they do not fit neatly into the scheme of things whereby humankind is at the unchallenged apex of a pyramid of power:

> 'Well, . . . a fox is a sort of special animal, isn't it? It's okay with sheep and cows and things because, like, they're just waiting there and not doing much. I don't mind the idea of that because, it's sort of like that's why they're there for us, isn't it? But a fox sort of like lives its own life. I mean, I can see why we have to hunt them to keep them down, but it wouldn't be right to eat them.'

Rodents likewise tend not to be eaten in western society. It is an established part of our historical mythology that a sure sign of a Desperate Situation, such as in a city under siege, is when its starving population resorts to eating rats. The *Monty Python* comic team play blackly upon this revulsion to provide a selection of recipes for preparing and cooking rats. As is so often the case with humour, however, their 'recipes' perhaps contain an element of truth in highlighting the torment inherent in animal cookery:

> *Rat soufflé*
>
> Make sure that the rat's squeals are not audible from the street, particularly in areas where the Anti-Soufflé League and similar do-gooders are out to persecute the innocent pleasures of the table. Anyway, cut the rat down and lay it

> on the chopping-board. Raise the chopper high above your head, with the steel glinting in the setting sun, and then bring it down – wham! – with a vivid crunch – straight across the taut neck of the terrified rodent, and make it into a soufflé.
>
> (Palin *et al.* 1973: 42)

We justify our revulsion on the usual scientific basis of hygiene and the threat of contagion ('you never know where they'd been, grubbing about and eating God-knows-what . . .') but, as Mary Douglas showed, our idea of dirt 'is compounded of two things, care for hygiene and respect for conventions' (Douglas 1966: 7). Douglas outlined how purity and defilement are reflections of systems of order and of contravention of that order (1966: 35) and that in any clear system of classification there may be intermediary or ambiguous cases which will tend to be treated with caution. It is significant that rodents are distinct in resisting clear categorisation, particularly with respect to their proximity to ourselves. In this case the uncertainty also relates to their residence: whether they are beasts of house or of field. Although rodents are not meant to be admitted to the human domain, they are renowned for their ability to invade. They even steal our food, scorning our best attempts to prevent them. They defy our classification as they defy our control, even within our own ordered human space:

> 'I'd rather have a spider around than a mouse.'
> *Q. 'Why?'*
> 'Because you can catch spiders and put them out! Mice run too fast and you can't catch them.'
> *Q. 'Why do you want to catch them?'*
> 'To put them out!'
> *Q. 'But why put them out?'*
> 'Because they run too fast and they scare me!'
> *Q. 'Why do they scare you?'*
> 'They go too fast! I suppose it's you can't control them. '

Each case dealt with in this chapter has shared the common attribute of being in some way ambiguous in our system of classifying the living world, particularly with respect to what we normally prefer to regard as the absolute distinction between the civilised human domain, and the wild residual category. Pets, primates, carnivores, rodents – all are avoided as food, since all are of uncertain identity. It may have been left to Mary Douglas formally to explain the correlation between purity and clearly ordered classification but it is a lesson which many of us are taught from an early age: 'That's what my mother told me: never eat anything unless you know what it is.'

10

THE JOY OF SEX

Man is the hunter; woman is his game.
The sleek and shining creature of the chase,
We hunt them for the beauty of their skins;
They love us for it and we ride them down.

(Tennyson, *The Princess* 1847)

Alex Comfort M.D. names his bestselling *Joy of Sex* (1974) after the classic American cookery Bible, *Joy of Cooking* (Rombauer and Rombauer Becker 1931). His subtitle – *A Gourmet Guide to Lovemaking* – continues the culinary metaphor. And he even orders his chapters like a menu: 'Starters', 'Main Courses', 'Sauces and Pickles', and so on. Comfort thus perpetuates 'a universal tendency to make ritual and verbal associations between eating and sexual intercourse' (Leach 1964: 53), explanations for which commonly focus upon some equivalence of our biological imperatives:

> Traditionally, hunger is seen as a basic drive for survival of the individual whilst sex is a basic drive for survival of the species. It might be expected that there could be found some parallels and interactions between these fundamental activities.
>
> (Fieldhouse 1986: 173)

'Natural' analogies between sex and eating (that both perpetuate life, that both may be pleasurable, and that both imply vulnerability by breaching normal bodily boundaries) may partly explain their connection. It is, perhaps, only to be expected that each should carry a multiplicity of meanings extending beyond its respective province, and even that they should be associated. But the parallels are nonetheless significant, since food selection reflects cultural conceptions, as do ideas about sexuality: 'What gender is, what men and women are, what sorts of relations do or should obtain between them – all of these notions do not simply reflect or elaborate upon biological "givens," but are largely products of social and cultural processes' (Ortner and Whitehead 1981: 1). The question is not why but how do we associate sex with eating.

One example is the notion of Man the Hunter, a figure who seems to stalk the world over. Collier and Rosaldo found 'unexpected regularities in the gender conceptions of several ("simple") societies', particularly that 'Man the Hunter, which we thought to be our myth, turned out to characterize their conception of maleness' (Collier and Rosaldo 1981: 275, 317). For the Masai youth the transition to adulthood is marked by his washing with the blood of a goat he has killed for sharing (Jacobs 1958: 7). A !Kung boy is considered eligible for marriage only after killing his first animal, and men are said to chase, kill, and eat women, just as they do animals; whilst 'femaleness negates hunting prowess' (Marshall 1976: 177, 270; Shostak 1983: 85). Similarly in north-east Peru, prestige gained through hunting brings men a definite reward: 'the possibility of gaining women as lovers and/or wives. It is a common feature that the Sharanahua share with all tropical forest hunters: The successful hunter is usually the winner in the competition for women' (Siskind 1973: 95–96). Many Spaniards equate bullfighting with male sexual prowess, the sword being the instrument of penetration. Amongst the Chipewyan of northern Canada, Sharp describes a rigid division of labour where hunting is an exclusively male preserve (Sharp 1981). And in British mythology, too, meat is a male preserve:

> I left the British Library and my research on some women of the 1890s whose feminist, working-class newspaper advocated meatless diets, and went through the cafeteria line in a restaurant nearby. Vegetarian food in hand, I descended to the basement. A painting of Henry VIII eating a steak and kidney pie greeted my gaze. On either side of the consuming Henry were portraits of his six wives and other women. However, they were not eating steak and kidney

> pie, or anything else made of meat. Catherine of Aragon held an apple in her hands. The Countess of Mar had a turnip, Anne Boleyn – red grapes, Anne of Cleves – a pear, Jane Seymour – blue grapes, Catherine Howard – a carrot, Catherine Parr – a cabbage.
>
> (Adams 1990: 26)

Countless such observations are made: men are routinely and ritually in a position of controllers, hunters, providers, and also primary consumers of meat, with first claim on available resources. Or to put it another way, meat is almost ubiquitously put to use as a medium through which men express their 'natural' control, of women as well as of animals.

The macho steak is perhaps the most visible manifestation of an idea that permeates the entire western food system: that meat (and especially red meat) is a quintessentially masculine food. This notion, for example, is central to a humorous book about gender-stereotypes which enjoyed brief celebrity in the early 1980s, and whose title, *Real Men Don't Eat Quiche,* still circulates as a catchphrase:

> In the restaurant, Real Men eat steak and chips . . . Real men never eat the compulsory sprig of watercress . . . In general, Real Men are not afraid of food. They do not flinch nervously at the sight of butter, white bread and refined sugar. They know that all proper meals are centred round meat, that yogurt is really milk that's gone off, muesli is some sort of chicken feed and salad is for rabbits.
>
> (Feirstein 1982: 72, 74)

Even in this whimsical context, it is instructive to note the association drawn between Real Man's high consumption of red meat and, according to the following introductory passage, His intrinsically destructive tendencies towards the natural environment:

> 'Real men don't eat quiche,' said Flex Crush, ordering a breakfast of steak, prime rib, six eggs, and a loaf of toast.
>
> We were sitting in the professional drivers' section of an all-night truckers' pit stop somewhere west of Tulsa on Interstate-44, discussing the plight of men in today's society. Flex, a 225-pound nuclear-waste driver who claims to be one

of the last Real Men in existence, was pensive:

> 'American men are all mixed up today,' he began, idly cleaning the 12-gauge shotgun that was sitting across his knees. Off in the distance, the sun was just beginning to rise over the tractor trailers in the parking lot.
>
> 'There was a time when this was a nation of Ernest Hemingways. *Real Men.* The kind of guys who could defoliate an entire forest to make a breakfast fire – and then go on to wipe out an endangered species hunting for lunch. But not anymore. We've become a nation of wimps. Pansies. Quiche eaters.'
>
> (Feirstein 1982: 8)

As the epitome of meat, a beef steak can send powerful sexual signals. The larger and juicier the piece of meat, the more red-blooded and virile the consumer should be supposed to be, and a steak by candlelight is a common prelude to seduction. Meat is widely reputed to inflame the lustful passions, particularly in men, the stimulation being generally of an animal rather than of an erotic kind (Twigg 1983: 24). It is reported, for example, that the captain of a slave ship, in the throes of evangelical conversion, stopped eating meat to prevent his lusting after female slaves (Cecil 1929: 118). Nineteenth- and even twentieth-century educationalists recommended a low meat diet for male adolescents to discourage masturbation (Miles 1904; Punch 1977). The association even turns up in academic literature. In surveying the nutritional superiority of meat, Harris notes that animal foods have been reported to be 'good-to-excellent sources of zinc, essential for male fertility' (1986: 36); again it is significant that it is particularly male potency which is conspicuously advantaged. Conversely, a male vegetarian can be a suspect figure, as a student recalls of a period living in a new community:

> 'It was really odd, they seemed to automatically assume that because I was vegetarian then I must be gay. I'm sure it was because of the thing about meat being a sort of virility symbol. And then of course, it wasn't helped by the fact that I was living in a house with a woman who wasn't my girlfriend – they couldn't really comprehend that either.'

A businessman expresses much the same idea, if as a joke:

> *Q. 'What do you feel about vegetarianism? Why is it becoming fairly popular?'*
>
> CLIVE 'Oh, they're just a bunch of cranks! And they should be lined up and shot, along with the poofs . . . [laughs]. Er, what do I think? I think they're human beings. They're perfectly entitled to do what they wish . . .'

Indeed, the same man projects similar values on to (quasi-human) pets, when his wife mentions a visit to her Italian aunt:

> PAT 'Yes, I mean, even the dogs were vegetarian, it was that . . . she was that bad about it. She had a strong belief in it rather. The dogs were perfectly healthy though; the two dogs were terrific, and were 100 per cent vegetarian. She did not let them have meat.'
>
> CLIVE 'God, their street-cred must have been absolutely the pits. Can you imagine going out with the lads, you know: "Come on, lets go and ravage a few cats" – "I'm sorry, I'm not into cats: any mushrooms we can ravage?" Woof!'

Man the hunter also survives in another somewhat impoverished form. Distinct parallels exist between the language of the meat system and a terminology that men use to describe women in pornographic and mainstream discourse. It is as if the one system of exploitation is modelled on the other, which is perhaps not far from the truth. The phenomenon is worth considering at some length since it is a principal metaphorical use of the concept of meat in the English language, developing upon its normal implication of strength, power, or challenge.

This usage has several components. One is men's description of women in terms of other animals: she might be a bat, beaver, bird, bitch, cat, chick, cow, dog, filly, lamb, minx, mount, mouse, nag, pussy, shrew, or vixen, for example. She might equally be a pet – a bunny-girl, a sex-kitten, or just pussy, for example – to be looked after and played with. As Susan Griffin points out:

> pornography is filled with associations between women and animals. We see a film in which women become animals, who are then trained with a whip. Juvenal tells us that a woman filled with sexual desire becomes 'more savage than a tigress that lost her cubs.' In *Hustler* magazine, a woman is photographed surrounded by the mounted heads of wild animals and animal skins. She opens her legs toward a live

> lion and touches her own breasts. Over the photograph we read that 'Lea' has shed 'the veneer of civilization for the honesty of wild animal passions.' 'The beast in her,' we are told, 'is unleashed.' Projecting even the mechanism of his own projection of her, the pornographer writes: 'She sees in wild creatures her own primitive lusts and desires, and she satisfies them with the uninhibited speed of a beast in heat.' And in the midst of several photographs of nude women who lie with their legs apart, revealing their vulvas, we find, in the same magazine, a photograph of a male lion, on his back, his legs, also, spread apart.
>
> (Griffin 1981: 24–25)

Another aspect is women's description in terms of hunting or farming: she can be ridden, bridled, and married to a groom. To make a 'catch', the man goes to a 'cattle-market' or out 'on the hunt'. Alternatively he may use his financial advantage to procure a woman 'on the game' – a name which combines affirmation of a woman's status as the man's plaything with evocation of her affinity to the 'sportsman's' quarry. Or perhaps the climax of his desires comes with her two-dimensional exposure in magazines such as *Rustler* (a name which means 'cattle thief'). In the first Reader's Letter in one such organ, Jack of Torquay relates an incident from his pastime of skeet shooting:

> Well, one Sunday I was shooting driven grouse . . . when to my horror I heard a scream from the woods, coming from the direction where I fired my last volley.
>
> I ran along to where I thought the sound had come from, and there, sitting on a log, was a young lady . . .
>
> Incredibly, she was unscathed – but for a piece of buckshot, lodged just beneath her skin, right between her breasts.

The only apparent method of removal is for Jack to suck out the offending buckshot, which seems to arouse the young woman:

> 'You can shag me if you want,' she said, dabbing the tiny trickle of blood that trickled down her cleavage with a tissue. 'My husband's been away for three weeks and I'm desperate!'.
>
> (*Parade* 83: 12)

The imagery could not be more clear, and is entirely consistent with the system of ideas whereby women are portrayed as Man the Hunter's willing prey. In similar vein, the so-called office wolf, or the wolf-whistles inflicted upon passing women by the stereotypical workman, are appropriately designated in view of the predatory relationships conveyed.

Another significant component of this language is the description of women as edible objects – a characteristic that is by no means confined to the English language, or to modern times:

> *The Dictionary of Historical Slang* lists several phrases colloquially popular since the C15th, some still in use, which evoke an image of woman as dead flesh, bloodily carved up, hacked at, minced by a butcher or cook, and eventually served up for male consumption. A *bit of meat* meant firstly sexual intercourse (from the male standpoint) and later a prostitute. *Fresh meat* was a prostitute new to the trade. *Hot meat*, sometimes *hot mutton* or *beef*, was used of a fast or loose woman, a prostitute, and for the vagina. *Raw meat* referred either to a prostitute naked in the sexual act or was a general term for any woman. *A meat house* was a brothel. A *meat market* was a term for a rendezvous of prostitutes, the female breasts and the vagina; it is used today by feminists to disparage beauty queen contests in which young women are encouraged to parade their bodies like cattle for sale . . . In the C20th the expression *a cut or slice off the joint* is a UK slang term used by men meaning to have intercourse with a woman.
>
> (Mills 1989: 155)

Whilst this vocabulary can be of edibility in general, the overwhelming majority of references relate women to meat in particular. A man might reckon her to be a tasty morsel, or delicious; he might fancy a nibble; or she might be sensually devoured. But ultimately what he is after is a bit of 'raw', 'juicy', or 'succulent' flesh to 'beef' in order to satisfy his sexual appetite, if he is 'hungry for love' in the words of a recent popular song. At its simplest, her body may just be described as meaty. One tabloid newspaper's 'exclusive' about a McDonalds' security man, for example, reports:

> *'Burger boss tells his new 15-stone bride:*
> *Lose any weight and I divorce you'*. . .
>
> Cuddly wife Ann has promised she won't fall even a quarter pounder because he loves his women beefy . . .
>
> 'I can't stand skinny dolly birds. Girls with meat on them make much better lovers'.
>
> (Parker 1987: 13)

The entire system operates as if women are perceived by men to be analogous to hunted, or else farmed, meat. Her body parts, portioned into the same names as the animals on a supermarket shelf – leg, thigh, rump, or breast – provide much of the innuendo-humour that fills the airwaves daily, whilst their innocent mention at table brings a blush to the faces of pubescent schoolboys. The humour or embarrassment takes effect by conjuring up images of women through the metaphor of meat, symbolically affirming their status in relation to men.

The sexual asymmetry is clear if we consider the situation transposed – as if it were men's bodies which were symbolically apportioned for consumption. With the exception of certain anomalous instances mentioned later, men are almost never so described in mainstream media, since masculine values still monopolise the cultural agenda. Men find little humour in having their thighs or breasts compared to those of, say, a chicken, as just another object of consumption. And, explicitly feminist jokes apart, women would tend not to deploy these particular associations. Men readily understand the joke about women, however. It amuses because the idea exists already in the mind.

Man the hunter illustrates a basic aspect of masculine thought, in patriarchal societies the world over, and not least in the reputedly modern west. It is as if in the traditional male cosmology, which emphasises the supreme value of prodigious power, women and nature have been regarded as analogous threats to his dominance.

Her position is defined relatively to his. As shown in Chapter 5, hunting is a primary characteristic whereby humanity is held to have first demonstrated – and for some still demonstrates – its civilised elevation above nature. And hunting is, archetypically, a male pursuit. Rosaldo and Atkinson, investigating the idea of 'Man the Hunter and Woman', believe it can be assumed that to some extent 'men and women are defined everywhere in relation to one another' (1975: 44), and Edwin Ardener argues that, as societies have always been observed to be to a greater or lesser extent patriarchal, the dominant indigenous model by which perhaps all cultures define and indeed create themselves, bounded against nature, is *inevitably* a male model, which effectively relegates important elements of female identity to wild nature:

> The objective basis of the symbolic distinction between nature and society . . . is a result of the problem of accommodating the two logical sets which classify human beings by different bodily structures: 'male'/'female'; with the two other sets: 'human'/'non-human'. It is, I have suggested, men who usually face this problem, and, because their model for *mankind* is based on that for *man*, their opposites, *women* and *non-mankind* (the wild), tend to be ambiguously placed . . . Women accept the implied symbolic content by equating *womankind* with the men's wild.
>
> (Ardener 1975: 14)

The nature–culture dichotomy is fundamental to recent (masculine) western traditions. It is one of many such dichotomies in our thought, just as most human societies use binary systems to classify their worlds (Needham 1973). Since theorising on the nature of our world has been traditionally the domain of middle- or upper-class European males, it is perhaps unsurprising that the following pairs should have come to be widely associated, as upper and lower, in both common and scientific thought (Birke 1986: 109–110):

culture	nature
man	woman
Europeans	non-Europeans
humans	animals
upper/middle class	working class

As Birke points out, whilst 'many such dichotomies might be demonstrated in Western thought, it is important to note that associations tend to be made on one side of the dichotomy; thus non-white people are associated with proximity to animals and to nature' (1986: 111) – or, as John Lennon puts it, 'woman is the nigger of the world'. Women are equated with nature and with animals; men are powerful, human, and civilised.

Thus we see that the argument proposed earlier, that human society has exhibited a consistent disposition to define itself as distinct from, and normally superior to, the rest of the natural world is now shown to be too general a statement, for different groups, including men and women, enter differentially into the debate, both in terms of ideas held, and in terms of access to the media of communication. Birke discusses at length the

> gender-related dichotomy . . . of nature versus culture, a distinction which has become central to Western ideas about the natural world and about humanity's place within it. 'Nature' is often regarded as somehow disorderly, chaotic and intractable; by contrast, our concept of 'culture' has come to include the capacity for human mastery over nature. Science, too, is implicitly part of that distinction, for it is science that has long promised to give us mastery over our environment, to force nature to yield up 'her' secrets. In the twentieth century, indeed, concepts such as 'progress' and 'culture' have become almost synonymous with those of science and technology.
>
> (Birke 1986: 107)

Thus civilisation is regularly usurped as a male prerogative whereas woman is placed closer to nature – an identity perhaps highlighted by her more obvious biological functions such as menstruation and child-bearing (Ortner 1982: 492). The psychoanalyst Guy-Gillet (1981), for example, argues that the blood spilled by the male butcher-priest in animal sacrifice has symbolic linkages with women's menstrual blood, the symbolic victims of man's devouring lust. In Victorian England, on the other hand, it was believed that women should not cure hams whilst menstruating. 'In 1878 a correspondence struck up in the *British Medical Journal* describing incidents when meat being rubbed with brine by ladies during those seasons had not taken the salt and turned bad' (Pullar 1970: 189). Whatever the clinical rationale, such beliefs serve to affirm women's subordinate profanity, contrasted to men's sacred gentility.

Griffin demonstrates the ubiquity of oppressive images not only in science and pornography, but in the art of Ramos, Tauzin, Milet, and others. She also notes the parallel equation of womanhood with brute nature in the work of such figures as Jung, who finds the African woman more female than 'civilized women'; Schopenhauer, who argues that women exist solely to propagate the species; Hegel, who writes that women cannot comprehend abstract ideas; and Augustine, who says that 'nothing brings the manly mind down from the heights more than a woman's caresses' (Griffin 1981: 26–27). Each example cited is, of course, a man, for it is men who dominate the agenda around which most of the official discourse of our society is oriented. Men have predominantly controlled education and information, law and its enforcement, marriage and alliance, production and distribution, science and technology. It is men who have set the agenda, and men who have ideologically established their own group as the axis of civilisation.

Nature and women have been characterised as twin threats to this supreme masculine power. On the one hand the untamed natural environment has been a physical challenge, whilst, in the cultural environment, women have evaded total masculine control. To the first is applied science and technology (which historically have been a male arena); to the other a range of legal, social, and economic constraints, and 'power-structured relationships' including the 'ultimate weapon' of rape (Brownmiller 1975), which are together described as 'sexual politics' (Millett 1977: 23). One such ideological constraint is the consigning of women to Mother Nature: to the category of the wild.

Men are also verbally and ritually associated with meat in some contexts, but normally in a quite different way than women. If men are referred to as meaty or beefy ('he's a real beefsteak') the equation is not so much with meat as a food, but is rather between the strength endowed by meat and his supposed sexual and physical potency. The muscularity that meat is reputed to endow is a popular masculine ideal. In the language of structuralism, it might be said that the conventional linguistic relationship of women to meat is metaphorical, whilst that of men is more often metonymical. In other words, men are meat in the sense that meat is full of power, whereas women are meat in the sense that it is consumed as a statement of power. The orthodox associations of masculine meatiness provides the basis for a semi-humorous newspaper item reporting a link-up between the publishers, Mills & Boon, and a meat firm which was packaging a free romantic novel with every 3lb of frozen sausage rolls sold:

> Whether the heroine of *Model of Deception* would find the thought of thawing sausage rolls attractive is another matter.
>
> In the words of the author, she preferred to discover that 'red hot blood had begun pouring through her veins. She couldn't be sure, of course . . . It must have been a trick, due to the sun perhaps, rather than Luke's overpowering masculinity.'
>
> All in all, a less prosaic defrosting process.
>
> (Cohen 1987: 1)

Presumably, had the heroine in question been offered proper red meat, she might have been better satisfied, gastronomically if not sexually. The symbolic statement made by mere sausage rolls, however, is somewhat incongruous, failing adequately to represent the object of her passionate-yet-innocent desires.

Only exceptionally, when normal male values are subverted in any case, can men be referred to as meat for consumption. In this somewhat subtle reference to meat-eating, the title serves to underline the abnormality of a situation in modern American society in which women can publicly command control of men's bodies by the use of money – the inversion of a privilege normally reserved for men:

> *Meat-eating in the States*
>
> When Sue Ellen raised money for good causes in *Dallas*, it was *never* like this.
>
> The 'exciting new way for single yuppie women and desirable bachelors to meet' has been launched on the innocent American public. The bodies of 39 eligible bachelors are being auctioned off this week. Former Redskins champion Babe Laugenberg, congressmen, and an ex-Mr Unique Physique are up for grabs, profits to go to a Washington charity for underprivileged children. The men will be available to the highest bidder – for an evening, or possibly for an overnight stay. Once sold, they may not say no.
>
> (Crawshaw 1987: 9)

In each of the following examples, similarly, the humour derives from violation of the principle that men are not normally meat to be eaten – and particularly not his penis, for which meat is a regular synonym:

I was eating a can of frankfurters and growing very weary of the demands from one of the onlookers for a share of my meal. When he finally asked what I was eating, I replied: 'Beef'. He then asked . . . 'What part of the animal are you eating?' To which I replied, 'Guess'. He muttered a contemptuous epithet, but stopped asking for a share.

(Chagnon 1968: 14)

I'll tell you a story which is beastly – but will make you laugh; – a young man at Ferrara detected his sister amusing herself with a Bologna Sausage – he said nothing – but perceiving the same Sausage presented at table – he got up – made a low bow – and exclaimed 'Vi riverisco mio Cognato' [I pay my respects to my brother-in-law].

(Byron, cited Bold 1980: 182)

Media historians will probably say this was the year when naked sex objects became respectable on British screens.

They will of course be talking about condoms. Your modest rubber johnnies went far in 1987. They were demonstrated on fingers and on plaster phalluses, they were recommended by Geldof – 'Don't forget, stick one of these on yer dick.' They were, hell's bells, advertised. They went all the way from derided options to medical necessities to comedy prop. There was Michael Palin on the David Letterman late night chat show stuffing one with fresh minced beef and slicing it salami-fashion. A Freudian frisson shook the land.

(Hebert 1987: 17)

The penis is quite often referred to as a man's sausage, usually in one of two contexts. Men tend to make such verbal 'links' exclusively in humour, where values can safely be transposed, as if to mitigate the potential ambiguity of the association. When women talk of men's sausages, however, the tone is more often mocking, chiding male arrogance by inverting the normal masculine system of ideas to invoke the notion of meat for consumption by women rather than by men, implying male vulnerability and, indeed, lack of virility or prestige. Rather than the potent red-blooded beefsteak as which the man might choose to see himself, he is derided as the possessor of nothing more than a mere sausage.

Women and men are conventionally allotted very different roles in the western food system. Women are generally expected to provide, cook, and serve food domestically, but men are still in some respects the ritual providers of meat, as if to fulfil their proverbial role as the hunter. In fact Jack Goody argues that:

> In human societies generally cooking is seen as part of women's role. That is not to deny that men may carry out other functions in the preparation of food. They are generally the killers of other animals (and of other men) as well as the butchers of domestic meat. Moreover, they often play a part in the roasting as distinct from the boiling of meat, in cooking in the fields or forest as distinct from the house, and in ritual as distinct from profane cooking.
>
> (Goody 1982: 71)

Men may be our traditional breadwinners but conventionally they do not actually buy the bread; they provide the resources with which it can be purchased. Even today, however, it may still be men who bring home the proverbial bacon: who notionally provide the meat. And the purchase of a proper piece of meat may still be seen as the model duty of the British man of the house, particularly for special occasions, ritually affirming his status. Just as with the !Kung and the Sharanahua, it is still our men who are the purported hunters and providers of meat, and women who 'forage' for the rest of the food. As one retired couple recount:

> HELEN 'Yes, that's the one thing I leave to John. At the weekend, you'll go out, won't you, and find us something nice for our Sunday lunch.'
>
> JOHN 'Well, not every Saturday. Often we'll just have a chicken or something to ourselves, but if the children are coming down then we'll usually try to see that there's something proper for them, so that's usually my job – to pick up a good looking roast or something – while Helen does the other bits.'
>
> HELEN 'And you usually find us something good, don't you?'

The fashionable garden or beach barbecue is another arena where the man is likely to take control, peculiarly evoking the 'fields and forest' mentioned by Goody. In the home the man may also be responsible for the roasting of a piece of meat, and particularly its carving – a significant activity for the male to have annexed, given its display of symbolic mastery whilst being devoid of any real requirement for skill or courage. It might indeed seem strange, if meat is the important cultural signal of human civilisation suggested, that women are permitted to take on the task of its cooking at all. But it is in routine and mundane situations that this typically obtains – it is the drudgery that is delegated to women; the exemplary prestige still accrues to the man.

Top chefs over the years have been almost exclusively male, and whilst those masters of cuisine may have many assistants male and female, the ancillaries are assigned the tasks of preparing vegetables, making sauces, assembling sweets and starters, washing dishes, cooking vegetables . . . but the chef generally takes charge of the *pièce de résistance:* the meat. At the pinnacle of food provision men still stand as figureheads.

Gender inequality is institutionalised in other ways too. Men have long been allocated an unequal proportion of available food, and particularly meat. In the working-class home at the beginning of this century, for example, men were consistently given the 'lion's share' (Pember Reeves 1979; Spring Rice 1981; Rowntree 1913; Co-operative Women's Guild 1978). Documenting this phenomenon, Kerr and Charles comment that:

> Women, on whose shoulders fell the responsibility for managing the limited household budget and for ensuring that the family was adequately fed, frequently did without food themselves in order that their husbands and children were less likely to go short. Men's food needs were often privileged because of the necessity of keeping them fit for waged work . . .
>
> (Kerr and Charles 1986: 116).

However, this pattern of maternal self-sacrifice has continued to be found in later poverty studies (Land 1969; Marsden 1969) and fathers' and children's needs are consistently put before those of women to this day (Kerr and Charles 1986: 116). Within this study of 150 families, for example, men generally ate more meat than either women or children, the only exceptions being low status meats such as sausages on which children were fed, and low status fish products, of which both women and children consumed more than men. Overall:

> meat consumption was generally high for all men in employment at the time of the study, regardless of the nature of their occupation, and these men almost without exception consumed meat more often than their wives while children almost always consumed meat less often than adults in these families. Very high consumption of meat was almost totally confined to men while very low meat consumption was confined to women and children.
>
> (Kerr and Charles 1986: 145)

Children, like women, traditionally are allocated a lesser amount of meat than grown men. Meat is the food of those who control the natural environment: who do what is regarded by real men as real work, as opposed to women's domestic chores. There is a tradition that less active members of society should eat less meat. Children are given special food, such as sausages, whilst their parents – especially father – enjoys real meat. The classic nursery diet is bland food, with little or no red meat. If animal flesh is given to children, it is in its milder forms: boiled or stewed, seldom roasted. Invalids too are commonly restricted to such diets, their constitutions reputedly being unable to deal with 'stronger' items. It has even been recommended that those engaged in bookish or sedentary occupations avoid excess consumption of red meat since it may stimulate the passions, with unhealthy frustration a probable consequence for those unable to dissipate through physical exercise the ardour it engenders (Twigg 1983: 25). In each case restrictions are couched in terms of digestion, but the groups whose intake of meat is limited are those least inclined to exert physical power over the natural world. Women, traditionally, are just such a group.

Women providing for families tend to suppress their own tastes and preferences, in order to come to a compromise for the entire family – compromises which, Kerr and Charles found, almost always meant the family eating whatever food the father finds acceptable, so that 'while women choose the food, their choice is largely dictated by men's food preferences. Most men were reported as being unadventurous eaters and it is clear that the importance men commonly attach to a plain and solid meal which includes plenty of meat severely restricts the variety which may be offered in the family diet' (Kerr and Charles 1986:

120–121), although this is not necessarily backed by forceful, or even overt, demands:

> In fact the women often commented that they attempted to gain help in decision-making about food purchase by asking their husbands for ideas but this was usually met with the response that 'anything' will do. Knowledge of men's preferences had been gained through a process of trial and error in the early stages of marriage and men had made their preferences clear through refusal or reluctance to eat food at the point it was presented to them. The strength of this sanction lay in the fact that it was likely to inspire concern and even guilt in the women as provider of food.
>
> (Kerr and Charles 1986: 121)

This inequality of distribution and choice is rationalised by women, as well as by men, with recourse to the supposedly 'natural' equation between males and activity; men are portrayed as more active than women, and boys than girls, and therefore require more meat. The male model of culture is by no means always opposed to women's ideas. Women, as Edwin Ardener suggests (1975: 14), accept much of the implied symbolic content of men's ideology, and incorporate it into their own. Women's valuations do, however, differ in detail considerably.

The portrayal of women, by men, as meat is an instance of the wider caricature of woman as animal, but it is an especially significant example. Just as meat is a sign par excellence of man's control of the natural world, so woman as meat has been a particularly effective statement of her supposedly wilder social role, and availability as a natural resource for men. The asymmetry of this subsystem of ideas is significant. Women are called meat as if assigned for men's consumption, but the converse does not occur: meat is not usually characterised as feminine. Nature provides an extensive and detailed model for the system of patriarchal control of women, but not on the whole vice versa.

This seems to undermine the argument that 'radical feminist theory exposed the inadequacy of all previous work on power, in failing to analyse the domination of women as a central form' (Eisenstein 1984: 131), and that 'sexual domination obtains nevertheless as perhaps the most pervasive ideology of our culture and provides its most fundamental concept of power' (Millett 1977: 25). This evidence rather suggests that domination of the natural world, as represented in the meat system, antecedes sexual domination, providing both a model and a metaphor for men's control.

Males dominate the official discourse of our society, and habitually associate women with meat, but these are not the only possible views. The implication of the inevitable dominance of the archetypical masculine ideology (which is, of course, a descriptive device, and is not necessarily the same as all *men's* views), rests upon a presumption of the necessary dominance of patriarchy, although there are signs that women's views have been becoming increasingly influential in society as a whole. Although women's models are partly imposed on them by the dominant male ideology, it would also 'be surprising if they bounded themselves against nature in the same way as men do' (Ardener 1975: 5). Women's views are related but distinct. Indeed, in recent years some women have sought to make a virtue of this previously disadvantageous status, by not rejecting but extolling their notionally closer relationship with nature. That this is now possible is partly due to progress in the relative statuses of women and men, and women's increasing economic independence, and partly due to significant changes in attitudes to the natural world outlined in Chapters 7 and 8.

Edwin Ardener looks to the lore of mothers and old women (old wives' tales) for the symbolic valuations that enact 'that female model of the world which has been lacking [in academic discourse], and which is different from the models of men in a particular dimension: the placing of the boundary between society and nature' (1975: 5), but the evidence is also readily available in the beliefs and actions of women in ordinary situations. The extent to which this is the making of a virtue of necessity from a situation imposed upon women by the dominant masculine ideology is open to debate; the fact remains. Women in western society do, in general, view themselves in a significantly different relationship with the non-human natural world than men: as more sympathetic, as less exploitative – which is often suggested as a probable reason for there being greater numbers of women vegetarians than men today.

Rosaldo and Atkinson (1975: 63) argue that 'celebration of motherhood and female sexuality implies a definition of womankind in terms of nature and biology; it traps women in their physical being, and thereby in the very general logic which declares them less capable of transcendence and cultural achievement than men'. Their study of the Ingolot demonstrates, however, that it is possible for women to transcend the implication of inferiority and achieve largely equal status, as has demonstrably been the direction if not the final position of gender relations in recent western history.

Masculine thought, it seems, has long been in the ascendant, but the twentieth-century upsurge of interest in both feminism and environmentalism suggests that attitudes to proper relationships in both spheres have recently undergone radical development. Women, like meat, have long been defined by men as natural, relative to dominant masculine culture. Women, like meat, to some extent retain that symbolic role. But our perception of each is changing, as the negative aspects of excessive male power in each sphere are increasingly perceived, by men as well as by women.

Part IV

MODERN MEATOLOGIES

11

ECONOMICS

Press releases announcing the results of the most comprehensive series of statistical surveys of vegetarianism in Britain cite only cost and health as significant influences. The ethical dimension is scarcely mentioned (Realeat surveys 1984–1990). The meat industry similarly stresses 'rational' considerations of nutrition and price in its promotion. For example, when the Meat and Livestock Commission launched a three-year marketing campaign, 'Persuading consumers that meat and meat products were nutritious, value-for-money foods [was] the ultimate aim of the plan' (*Butcher & Processor* 1987b: 5). In each case the implication is that consumers choose by evaluating clear criteria such as cost and healthiness. It is simply assumed that meat is highly valued, as shown by its regular and long-standing use as an *indicator* of welfare and economic development:

> The better-paid workers, especially those in whose families every member is able to earn something, have good food as long as this state of things lasts; meat daily, and bacon and cheese for supper. Where wages are less, meat is used only two or three times a week, and the proportion of bread and potatoes increases. Descending gradually, we find the animal food reduced to a small piece of bacon cut up with the potatoes; lower still, even this disappears . . .
>
> (Engels 1844: 372)

Nakornthab (1986) uses meat eating similarly as a primary measure of Thailands' urban development, alongside use of electricity, tax revenue, and car ownership, and Shields (1986) takes rising meat consumption as a 'clear illustration' of increasing prosperity in the Iraqi city of Mosul in the nineteenth century. David Riches, similarly, disputes Sahlins's conception of hunter-gatherers as the 'original affluent society' (Riches 1982) on the grounds that in many such societies 'the consumption of meat is very highly regarded [yet] constitutes as little as 20 per cent of food intake', from which he concludes that their meat intake is, for them, inadequate. Its securing, he holds, is inhibited by cultural factors such as gambling which restrict time-consuming hunting (Riches 1982: 216). However, to dispute a people's well-being on the grounds that they prefer to allocate time to one leisure activity (gambling) than to an arduous activity (hunting) required to obtain extra quantities of a particular good, is hardly supportable. That this does not reflect how the particular people concerned think hardly enters into the matter – an attitude Riches justifies on the grounds that interpretation of behaviour is the anthropologist's prerogative (Riches 1982: 6) – a defensible argument taken to indefensible lengths. Riches falls into the trap of seeing meat as inherently desirable in unlimited amounts, presumably because we esteem it so. This example has a wider relevance than to the subsistence-economy peoples discussed by Riches, for the same attitude occurs in standard economic analyses of 'modern' societies.

To accept unquestioningly that meat is desirable or prestigious, as if these attributes were in some way *inherent* in the substance itself, is to oversimplify the range of ideas that meat supports. Its economics cannot properly be studied as a disembodied quantity. Meat's value reflects its appraisal by individuals within a culture, and it is the *reasons* for that appraisal which must be investigated, to which end analysis of abstract statistics may be little more than a circular diversion.

This is not to suggest that economists' calculations, such as of the 'price-elasticity' (the relationship of changes in price to changes in demand) and 'cross-price effects' (the effect of the changes in price of one meat on demand for another) of the different meats, are not valid analytical procedures (e.g. Peters *et al.* 1983). But these cannot be more than 'snapshots' of idealised representations of aggregate behaviour in any society. Percy Cohen notes that:

> it is generally recognized that, although economic analysis can usually explain changes in certain values, such as wage-

> rates, it can explain neither the actual values themselves nor, therefore, why some values are much lower than others, since these have their origins in certain customary arrangements of the past.
>
> (Cohen 1967: 102–103)

Since Mauss published his *Essai sur le don [The Gift]* in 1925, substantivist economists have repeatedly shown that attempts to consider exchange of goods in traditional societies outwith their social setting are badly misguided, and that exchange can only be understood as one aspect of the system of social interaction in general (e.g. Dalton 1961). Even in our own nominally secular society, economic exchange is social exchange, and formal economic concepts, indiscriminately applied, may be just as inappropriate as they are seriously misleading when imposed on other cultures. As Marshall Sahlins puts it: 'What is finally distinctive of Western civilization is the mode of symbolic production, this very disguise in the form of a growing GNP of the process by which symbolic value is created' (Sahlins 1976: 220).

Economics is nonetheless still commonly represented as if it were an independent force in our affairs, governing habits rather than reflecting them. One piece of research investigating the factors governing demand for animal foods in the US, for example, concludes that households 'appear willing to pay for both nutrient and nonnutrient components of food items. The demand for these components was related to the socioeconomic characteristics of the households and to their incomes' (Hager 1985: abstract). In other words, it was found that the affluent tended to buy different sorts of meat from those bought by poorer people – not, one might think, the most remarkable piece of research. The significant point, however, is that the only nonnutritional aspects of value considered were such things as the way the meat is cut and presented. Intrinsic value is assumed.

Such intractably static conceptions are, however, poorly equipped to interpret the rapid and basic changes in the traditional perception of meat which today bedevil the meat industry. Meat has been in decline at a time of rising affluence. Spending on red meat in particular, as a proportion of household income, is projected to continue to fall, leading the British Meat and Livestock Commission correctly to reason that 'as lifestyle and priorities change, so too do eating habits' (*Meat*

Trades Journal 1989d: 20). Indeed, measured as a proportion of average income, the 'real' price of meat in Britain has actually fallen considerably in recent years:

Figure 6 Length of working time necessary to pay for 1lb of rump steak, 1971–1988

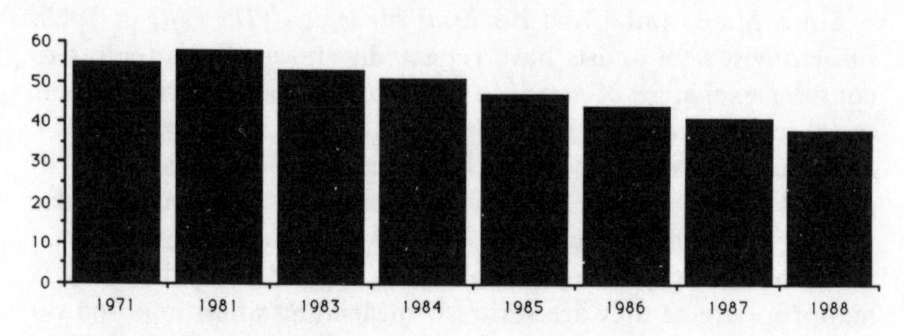

Source: Social Trends 1990: 103

To individual consumers it may well appear that the high cost of meat is the reason for their unwillingness to eat it. To the market researcher, who enquires about the motive for their abstention, it may likewise seem a reasonable explanation. But only by delving deeper into the ideas associated with that food, and identifying the beliefs which influence their acceptance by particular individuals, can reasons be established that lead some people to value meat less highly than others, and so be unwilling to pay its price. High cost, in any case, might lead people to eat less meat, but not to avoid it altogether:

> *Q. 'So why don't you eat meat?'*
>
> 'Oh, I can't afford it. My boyfriend and I are both living on grants and we just can't manage it.'
>
> *Q. 'Is that really the only reason?'*
>
> 'Yes. Absolutely. We just don't have the money. It's far too expensive.'
>
> *Q. 'So you still enjoy eating it if you've been invited to dinner by friends or something then?*
>
> 'Well, no, I still prefer not to really.'
>
> *Q. 'What, even if you're not paying?'*
>
> 'Well, yes. I don't really know why. I just prefer not to. I know it's silly, but we'll usually ask if we can have something else.'
>
> *Q. 'You must have some idea why, surely? Give me a clue?'*

> 'I don't know. I just don't like the taste.'
>
> *Q. 'Do you mean you've grown out of liking it after not eating it for so long?'*
>
> 'No, I don't think it's that. I suppose . . . it's something to do with not liking the thought of . . . I don't know. Just not liking the idea of the animal being . . . killed . . . so that I can eat it. It's horrible.'

Price alone could not explain the existence of vegetarianism. Another sign that changing buying habits do not merely reflect changing prices is that many people report a willingness to pay substantial premiums for humanely produced foodstuffs, despite acknowledging that there may be little tangible difference in the product purchased. Some individuals may be aware of, for example, the moral considerations which they allow to modify their rational economic choices. For others the ideological judgement involved may be so taken for granted that it is not recognised at all. Nonetheless, the *meaning* of a commodity plays a key role in all transactions, since an individual's beliefs constitute the essential context of their economic behaviour:

> 'On the question of cost too . . . My sister, who's vegetarian now – when she was a student it came about initially I think because they couldn't afford to buy meat, but then . . . I mean, they're both ecologists as well.'

Capitalist economics embodies a multitude of value-assumptions which normally go unnoticed, still less challenged. Our restricted view of economics is part of a modern mythology of rationality and objectivity, as if its study can be conducted in value-free isolation from social considerations. Incantations replete with economic jargon are ritually uttered by politician and pundit alike as both explanation and solution to almost any class of problem, and whoever dares defy its reputedly immutable laws risks intellectual perdition. But the market mechanism, ostensibly divorced from the social sphere, is itself a cultural artifact that reflects a particular point of view. It posits a split between the human spirit and its material setting, between the world of belief and the world of action, between dreams and reality, and this is itself illusory. Karl Polanyi charts the development of this peculiar brand of economics alongside industrialisation, and the consequent confusion of its narrow frame of reference with that of the broader human economy of social activities and institutions:

> Accordingly, there was a market price for the use of labor power, called wages, and a market price for the use of land, called rent. Labor and land were provided with markets of their own, similar to those of the proper commodities produced with their help.
>
> The true scope of such a step can be gauged if we remember that labor is only another name for man, and land for nature. The commodity fiction handed over the fate of man and nature to the play of an automaton that ran in its own grooves and was governed by its own laws.
>
> (Polanyi 1977: 10–11)

Polanyi argues that it is a grave error to equate 'the human economy in general with its market form', a distinction repeatedly obliterated by the economic *Zeitgeist* (Polanyi 1977: 6). To assume that the economic laws by which we operate today are in any sense eternal or absolute is positively misleading, which is not to say that such an assumption is rare. One important effect of the confusion of the market economy with its broadly based social counterpart, Polanyi suggests, is an excessive emphasis on material motivations for human activities to the virtual exclusion of all other incentives. We thus warp our understanding of ourselves and our society.

Reducing human activities, needs, ideals, and aspirations to the manageable level of unitary transactions may be justifiable as a device for certain clearly delineated purposes. But the extent to which such abstract analysis is widely accepted as a 'true' measure of individual interests should be, at the very least, contentious. All the more surprising then that this sort of approach can still pass as largely unchallenged orthodoxy. For example, one might properly regard with considerable caution a system in which every penny spent on constructing weapons of destruction, on combating rising crime, on dealing with increasing incidence of mental and physical disease, and on mitigating the worst consequences of pollution, is automatically included as a positive benefit in calculating the principal index of social welfare – gross national product. When that system is intrinsically unable to calculate the diverse range of individual and communal considerations and values that contribute to every economic transaction it is arguably time for its assumed applicability to be severely curtailed. The approach, according to some, presents an 'awesome catalogue of failure and misconception':

> With their money-based indicators and targets, oriented almost exclusively towards the formal economy, economists have consistently misread the situation and many of their prescriptions and remedies have actually caused it to deteriorate. Among other failings, these indicators are inclined to confuse costs and benefits, leave social and environmental factors out of account, and ignore the informal economy altogether as a source of work and wealth.
>
> (Henderson *et al.* 1986: 38)

But this ideology still holds sway in inappropriate contexts. Marvin Harris, for example, holds that 'there are generally good and sufficient practical reasons for why people do what they do, and food is no exception'. He argues that considering anything but material explanations tends only to prevent 'people from understanding the causes of their social life' (Harris 1986: 14; 1975: vii). His analyses accordingly involve straightforward forces that 'cause' us to behave as we do, such as a rational nutritional motive for cannibalism (the existence of which he does not doubt). Insofar as Harris is a product of modern western culture his views are perhaps not surprising, but the irony of his claim that 'people are taught to value elaborate "spiritualized" explanations of cultural phenomena more than down-to-earth material ones' (1975: vii) is that, on the contrary, it is normally only that which can be directly observed and measured that is today afforded real credibility. His theories offer alluringly scientistic explanations to a society that has become used to being told that anything else is practically an irrelevance. The interesting thing about Harris's position is his determined denial of a social component to social activities.

Time and again, we effectively marginalise non-monetary aspects of human interaction and regard only rational, money-based activities as real, and thence only the economic system as real society. 'Formal' economics is instrumental in shaping our world view, with social and political premises, and social and political effects. It is part of the masculine ethos that has been in the ascendant in western culture throughout recent history, stressing mechanism, reductionism, competition, hierarchy, and dominion. According to Bertrand Russell, the pivotal emphasis given to the competitive element ('survival of the fittest') in evolutionary theory by Darwin, and particularly by his followers and interpreters – which is still regularly invoked to justify exploitative practices in political and commercial affairs as well as in the meat supply system – largely mirrors the thrust of the same political ideology:

> Darwinism was an application to the whole of animal and vegetable life of Malthus's theory of population, which was an integral part of the politics and economics of the Benthamites – a global free competition in which victory went to the animals that most resembled successful capitalists.
>
> (Russell 1946: 808)

Significantly it is commonly those in positions of power, who determine the agenda upon which debate is based, who stand to gain most from the perpetuation of this ideology. Those in authority have for many years propagated an influential set of beliefs that are consistent in their view of the proper relations between the strong and the weak, whether the subjects be human or animal, which tend to regard power as the lawful preserve of the powerful, and in which dominion over the natural world is an integral component. Indeed, it is interesting to hear a speculative correlation drawn between societies, such as in Polynesia, which live by vegetable growing and whose political system was generally non-hierarchical, and those whose management of herds may have engendered more authoritarian politics (Thomas 1983: 46, discussing Haudricourt 1962). The ideology of domination of the natural world, which emphasises meat eating, is regularly associated with political beliefs with such ideals as authority, tradition, laissez-faire economics, and libertarianism. Conversely, it is fitting that vegetarianism, which tends to advocate less exploitative relationships with animals, should traditionally have been linked to more egalitarian ideologies. George Orwell, for example, suggests that:

> The typical Socialist is not, as tremulous old ladies imagine, a ferocious-looking working man with greasy overalls and a raucous voice. He is . . . more typically, a prim little man with a white-collar job, usually a secret teetotaller and often with vegetarian leanings . . .
>
> (Orwell 1937: 152)

Market forces are almost religiously believed in by many influential people today – revered as the ultimate arbiter of what is right and proper, absolving the individual of the burden of personal responsibility. We accept our culture's values as natural and incorporate them into our own beliefs, effectively perpetuating the status quo. Economics, in this way, has come to enjoy a uniquely mythological status in recent thought. But we should not permit collective fixation upon a tool of analysis to lead it to become our focus. Enchantment by such a simplistic explanation for the complexities of human behaviour should not dull sensitivity to alternative ways of understanding.

Meat eating, like any consumption, is a manifest expression of personal ideas in a cultural context, and it is to these *ideas* we must look for the source of observed value, since (in accordance with the economists' concept of opportunity cost) if we choose to invest time, money, or any other currency, in obtaining, preparing, offering, or eating meat, then we do so in preference to alternative ways of allocating those resources – for other foods, other activities, or other possibilities such as charity or saving. It is necessary to ask *why* the food is highly esteemed, and the answer can only be found in the ideas communicated by individuals in society. In the case of this particular commodity, our prevailing ideology with regard to the natural world substantially informs our transactions – dictating whether we place high value, low value, or even negative value, on the flesh of other animals as an object of consumption.

Although nothing is intrinsically luxurious, meat may justifiably be said to be a Natural Symbol of high value. It is an obvious choice of focus for our esteem in view of the spirit in which, traditionally, we have approached the world.

Meat is the flesh of what were once living animals; it is destined for our physical consumption. This makes it an exceptionally well suited exemplification of our ability to control and vanquish the non-human world – a goal, as shown in Chapters 5 and 6, upon which we have (perhaps with good reason) placed great emphasis. This is not to say that meat will *inevitably* hold elevated social status, only that it is likely to be viewed positively so long as our ability to control the wild is highly valued. This explanation is complementary to nutritional motives for meat eating, which are not in dispute, but which fail to account for the pre-eminence of meat.

Even within the narrow remit of formal economics it has long been recognised that objects are not difficult to acquire because they are valuable, 'but we call those objects valuable that resist our desire to possess them' (Simmel 1907: 67). The natural environment resists our desire to possess and control it. To the many human cultures which have striven to establish their identity apart from and above the rest of nature the consumption of animal meat is an eminently suitable choice to represent power, achievement, prestige, civilisation: humanity. Meat is partly valued *because* it is expensive to produce in terms of effort and of environmental cost, not in spite of it, for much the same reason as led Lévi-Strauss to note that roasting is the most prestigious method of cooking meat in many societies, since it incurs the greatest wastage (1966). Marvin Harris elevates this aspect of meat's value, to the exclusion of all others, in postulating that in agricultural societies 'animal foods are especially good to eat nutritionally, but they are also especially hard to produce. Animal foods get their symbolic power from this combination of utility and scarcity' (1986: 22). This insubstantial conjecture obscures, however, the widely varying values attached to meat, including the many individuals, and societies, who prefer a vegetarian diet. Meat's economic primacy is conditioned at an altogether different level of thought: by what meat represents to our society – control of the natural world and everything in it.

Time and again we encounter cases of meat consumption as a motif representing power and affluence, or of its absence representing deprivation. When the presenter of the BBC Radio 4 morning news programme, *Today*, wishes to commit the government minister for social security to a reasonable definition of poverty, he asks 'Would you agree that someone is poor if they cannot afford to buy meat regularly?' (12 May 1989). When a local newspaper wishes to feature the hardship suffered by students on diminishing funding it publishes a photograph of one sitting huddled in the December chill, with the caption 'Lucy: a treat is drooling over steaks in the supermarket' (Smith 1988: 14). When Khrushchev wished to boost the status of the Soviet Union's communist system one of his principal goals is said to have been to overtake the United States in meat production (Voinovich 1987: 13). And when Marvin Harris wishes to score political points for his American homeland, whilst ostensibly explaining his pet theory of 'meat hunger', he proselytises:

> Picture a line of people dressed in shabby raincoats, umbrellas in one hand and an assortment of plastic bags and briefcases in the other. As they shuffle forward in the grey dawn, the ones up front grudgingly make room for women who are pregnant or carrying infants. Those behind grumble and make jokes about pillows stuffed under dresses and babies borrowed for the morning. One woman in a knit cap explains: 'Nothing has gone up in price at this stand because there's nothing here anyway.' The Polish people are beginning their daily hunt for meat.
>
> (Harris 1986: 19)

Consistent with this are the many and various associations made between the monetary system and the animal foods system. Thus meat is not only an expensive food financially; the culinary term 'rich' is extensively applied to flavours deriving from animal products. Meat, and possession of live animals, have historically signified wealth and strength. In the *Iliad* (vi, 234), Homer ridicules Elancus for exchanging his golden armour worth 100 oxen for Diomedes' bronze armour worth only 9 oxen. Control of animal muscle both begets and denotes economic 'muscle'. Thus the word pecuniary derives from the Latin pecus, meaning cattle, and chattels, meaning moveable property, derives from the same old French word as cattle; as Jonathan Culler points out, linguistically the English word cattle at one time meant 'property in general, then gradually came to be restricted to four-footed property only . . . before it finally attained its modern sense of domesticated bovines' (Culler 1976: 22).

Meat has been a potent symbol of power over the wilderness, perhaps particularly historically whilst it has been difficult and costly to obtain and to maintain. The supply of animals, and thus of meat, has tended therefore to be controlled primarily by the wealthier, the more skilled, the more powerful, the central actors in the human drama . . . those who accredit themselves as the more civilised of people. Its provision and consumption has in turn been used to demonstrate the supposed affluence and sophistication of those who command its supply. The circle of meaning is complete.

12

HEALTH

Protein. Its very name means 'primary'. It forms 'an important part of all living organisms, and the essential nitrogenous constituents of the food of animals' (*OED*). The primacy of animal protein as a source of human food was scientific orthodoxy until well into the twentieth century, and the idea of nourishment remains the most heavily promoted explicit value used by meat suppliers. Much effort goes into informing the public that meat is an indispensable part of a healthy diet, and throughout modern society belief in the essential vitality of high protein intake – usually meaning meat – remains widespread:

> 'With an adult it's not so bad, but it's with growing children they say there can be a problem with that form of diet, because they've got to eat so much more to get the same protein levels – i.e. to get that high protein level it's not going to be healthy for the child to eat that amount that's going to be required, if you see what I mean. Because with a high protein food, such as meat, the child is feeling satisfied and getting a good high protein ration. I believe that's right . . .'
>
> 'I don't think I could become completely vegetarian. I think I'd find that difficult. Because then you have to work out your protein balance, and make sure you're getting enough soya beans and things.'
>
> 'You've got to be joking. I'd die!'

Popular belief in the need for large quantities of protein in the human diet can be traced back to the work of Baron Justus von Liebig, an eminent nineteenth-century German chemist. In *Animal Chemistry* (1846) and *Researches on the Chemistry of Food* (1847),

Liebig glorified meat as the essential source of material to replenish muscular strength. He thus 'gave new scientific status to current notions that animal food was somehow more nutritious than mere vegetables, and his prestige soon endowed meat with near-magical properties' (Griggs 1986: 15). Liebig endorsed the erroneous nineteenth-century view that muscle was destroyed by exercise, and could be replaced only by more protein, or in other words meat. This despite knowing that most primates and other mammals subsist on little or no meat, and that countless humans also live healthily on vegetable diets. (And, indeed, true carnivores consume raw flesh, innards and bone, and so obtain the necessary nutrients that our favoured diet of cooked muscle lacks.)

There are several points worth making about Liebig and the origins of what has become known as the Protein Myth. The first is that as one of the foremost theoretical chemists of his day and, amongst other things, a pioneer of chemical fertilisers, Liebig was personally committed to the scientific enterprise of human dominion over nature, and therefore can hardly be said to have been impartial in his attitudes. That he should have regarded meat highly himself is quite unsurprising; reducing 'meat' to 'protein', for scientific respectability, was merely his trade.

It is particularly interesting that Liebig should have propounded muscle fibre as carrier of the putative physical power, since the notion that food can somehow convey the physical or spiritual qualities of its source is widespread. As Simoons notes whilst discussing avoidances, 'Primitive man is greatly concerned about the flesh food he eats: in consuming a fellow creature he exposes himself to all sorts of physical and spiritual influences' (Simoons 1967: 117). The notion that 'you are what you eat' is common to modern Buddhists as it was to fourteenth-century Albigensians, who believed that 'to eat animals was to interfere with metempsychosis, the vast circulation of souls between birds, mammals and men' (Le Roy Ladurie 1978: 9). A similar notion was noted two centuries ago amongst American Indians who believed that a person who eats venison is swifter and wiser than one who eats 'the flesh of the clumsy bear, or helpless dunghill fowls, the slow-footed tame cattle, or the heavy wallowing swine' (Adair 1775: 113). Indeed, to this day, *Napoleon* brandy is reputed to contain infinitesimal proportions of the original spirit in which the hero's body was preserved for return to his native land; even in homeopathic doses, the drink is rumoured to fortify the drinker with something of his essence.

Simoons's ethnocentric attribution of such beliefs only to non-western peoples should not obscure our own similar proclivity. Liebig's enduringly influential ideas entail a variation on the notion that muscle begets muscle, their presentation in scientific language notwithstanding. It is doubly curious that a scientist whose work on plant nutrition was revolutionary in refuting the older theory that life could only proceed from life by showing combinations of pure minerals to support growth should wrongly assume human muscle to proceed only from animal muscle. Despite his theoretical positivism, Liebig uniquely exempts humans from the laws that govern the rest of nature – staunchly supporting the doctrine of human uniqueness.

But Liebig's personal ideology is perhaps not as significant as the academic acclaim, and enormous popular influence, that greeted his theories. 'Not only chemists and physiologists read and discussed his work, but amateur scientists, journalists, and even housewives became familiar with Liebig's ideas and his classification of foods. "Carbonaceous" and "nitrogenous" foods were as widely known in the 1850s as calories and vitamins are now' (McLaughlin 1978: 65). His popularity has a straightforward explanation: Liebig's theories were accepted almost uncritically partly because of his established reputation but also because middle-class Victorian biologists were themselves great meat-eaters by choice and, like the public, were pleased to learn that a high protein diet was scientifically approved (McLaughlin 1978: 67; Rivers 1981: 8).

Whilst Liebig was singularly instrumental in endowing meat with its modern, reductionist *alter ego,* protein, he was certainly not alone; animal flesh has long and widely been seen as embodying strength and vigour more than any other food, and scientific research has regularly substantiated this. Maimonides, for example, twelfth-century personal physician to Saladin, stressed the importance of diet as medical treatment, with mutton, chicken and gamebirds, and also wholemeal bread, heading his list of good foods (Griggs 1986: 3).

Particularly between the eighteenth and the early twentieth centuries such authoritative advice appeared in abundance – a period during which, as already noted, meat consumption was in any case rapidly rising. In 1833 William Beaumont, a US army surgeon, published descriptions of the stomach's operations, based on observations of a fur-trapper's unhealed wound, and concluded that 'generally speaking, vegetable aliment requires more time, and probably greater powers of the gastric organs, than animal' (Griggs 1986: 6). Two years later Thomas Graham agreed that animal food is 'no doubt, more allied to our nature, and more easily assimilated to our nourishment' whilst vegetables are 'digested with more difficulty', explaining that in 'the stomach, vegetable food always shows a tendency to ascendency, while the latter in moderation is almost never felt... There are few who subsist entirely upon vegetables, and of these few, the constitutions are generally feeble, sickly, and weak' (Graham 1835: 134–135). Science established what its practitioners and audience already knew – that meat was essential food. Its power was directly linked with animal strength, through a metaphorical, as much as any nutritional, connection.

To this day, some argue that meat consumption is physiologically embedded in our constitution. Marvin Harris, for example, still sells books on the premise that our 'species-given physiology and digestive processes predispose us to learn to prefer animal foods. We and our primate cousins pay special attention to foods of animal origin because such foods have special characteristics which make them exceptionally nutritious', he says (Harris 1986: 31). The facts, however, are not on his side. Other primates in the wild, for instance, pay considerably less attention to meat than we do. Harris can only allude to vaguely defined 'characteristics' since science has largely deserted him.

Harris's muddled claims may be symptomatic of a trait reported by many individuals, vegetarian and meat eating. They believe that eating animal flesh can engender physical changes, perhaps in the stomach's enzyme balance, which generate a sort of addiction that can be outgrown after a few days, weeks, or months of abstention. Biological habituation might, indeed, cast light on the phenomenon of meat-hunger whereby some people feel unsatisfied by a meal without animal flesh, perhaps intensifying their conviction of the need for it nutritionally. It could also help explain certain bodily characteristics reported in association with meat eating:

> 'Meat smells! Urgh. I never noticed before, but I notice now. I can smell it on people if they've been eating meat. I can tell. I mean, I have to be really close to them. But if Ian has been eating meat I can smell it... on his skin. It's

horrible! It's really vile. I never used to be able to, but yeah, I can tell when he's been eating meat.'

'Oh yes, it's definitely true. You can always tell a meat eater because their farts are so indescribably smelly! You can't help but notice it if you mix mostly with veggies, because . . . well, it's true that if someone eats masses and masses of beans, for example, then they might fart quite a bit . . . but the thing is that it doesn't actually smell all that objectionable. Whereas if you get a meat eater pumping away next to you, you'd better get out of the way! I mean, it's just totally different. They're, like . . . really bitterly rancid or something. You can smell the death erupting from inside them – literally.'

This latter trait also suggests an interesting link with the widespread association of meat with aggression. As a habitual vegetarian travelling through African rainforest reports in a letter home, those of us who eat animal flesh may actually send subliminal signals describing our nature:

'Recently I have occasionally noticed a strange smell round the place and looked at people near me. Suddenly I realised it was me, and the reason is because I have been eating meat since coming to Africa. So you don't have to go very deep to see why animals and birds can know whether you are to be trusted or not!'

Such evidence is, as yet, largely anecdotal; nor has even a short-term physiological basis for 'meat-hunger' – as distinct from the obvious mental yearning – been reliably established. But perhaps popular testimony should not be entirely discounted, particularly when it might be the harbinger of tomorrow's conventional wisdom. After all, time and again, nutritional and biological science has seemingly done little but authenticate its audience's predispositions, and tastes are clearly changing.

Just as scientific research long authenticated popular belief in our need for meat, so in the twentieth century, as more individuals were in any case reducing their flesh consumption, reports from the citadels of the scientific establishment began to endow previously marginal beliefs with new respectability. Even within living memory, orthodox nutritional standards have been modified considerably:

> [In] 1948, when protein was in the ascendant, the National Research Council of the United States recommended a protein intake for small children that was almost precisely twice what was recommended by Britain's Department of Health and Social Security in 1969; and . . . the amount recommended by the FAO and the World Health Organisation . . . is now about half that DHSS figure. The practical significance of this is that it would have been hard to meet the NRC's 1948 recommendation without heavy reliance on animal products . . . To meet present requirements, you need use animal products only to supplement plant protein – or indeed, as vegans demonstrate, you do not require them as a protein source at all.
>
> (Tudge 1985: 112)

Twentieth-century scientific reports have increasingly concurred that meat is not after all an absolute prerequisite to a healthy diet, finding, for example, that 'it is difficult to obtain a mixed vegetable diet which will produce an appreciable loss of body protein without resorting to high levels of sugars, jams, and jellies, and other essentially protein-free foods' (Hegsted *et al.* 1955: 555). A major review of the scientific literature on vegetarian or near-vegetarian diets published in 1964 concluded that a 'reasonably chosen plant diet, supplemented with a fair amount of dairy products, with or without eggs, is apparently adequate for every nutritional requirement of all age groups' (Hardinge and Crooks 1964: 537). Indeed, the American Dietetic Association now recognises that 'most of mankind for much of human history has subsisted on near-vegetarian diets' (Brody 1981: 438).

There has been no shortage of recent research suggesting high meat consumption to be not just unnecessary, but unhealthy in many ways – some intrinsic, some due to production methods. Peter Cox, for example, takes a campaigning stance to describing the gamut of concerns. These include chemical hormone residues in meat, whose use he suggests is probably more widespread on the black market than it ever was when legal (Cox 1986: 24) – a concern reflected even in the *Meat Trades Journal* which recently headlined its front page with the news that 'Farmers flout loophole in hormone ban' (1989b: 1), reporting routine discoveries of hormone-treated cattle at abattoirs, with inspectors powerless to prevent the trade. Cox also documents the evidence of links between meat, fat, and protein consumption and various cancers; meat's alleged indictment as a significant factor in coronary heart disease; and also increasing antibiotic immunity, allergic reactions, diabetes, hypertension, gallstones – not to mention a general reduction in fitness (1986: 35–67, 82–102).

The vast majority of food poisoning is attributable to meat and animal products, reports of which trebled since 1982 (Ehrlichman 1990: 3), and concern for which has increasingly disrupted sales; cooked meat sales in Cheshire and North Wales, for example, slumped by half after salmonella poisoning was linked to pork leg (*Meat Trades Journal* 1989f: 3). Fewer than 10 per cent of UK slaughterhouses are hygienic enough to be allowed to export meat within Europe, with inspectors complaining of inadequate sterilisation and ruptured intestines smearing meat with faeces. Infections transmitted by meat include salmonella, campylobacter, tape worms and other parasites, listeria, toxoplasmosis, and chlamydiosis (Ehrlichman 1990: 3).

And more serious threats continue to be discovered. The trade press has reported that 'British beef may be the source of a rare organism that can cause potentially fatal kidney failure', reporting the discovery by Sheffield public health researchers of an associated verotoxin producing *E. coli* (*Meat Trades Journal* 1989c: 15). A study in Sweden suggested that the breast milk of women on lacto-vegetarian diets contained the lowest levels of DDT compounds, PCBs, and other environmental contaminants (Noren 1987). Animal foods are regularly shown to contain higher residues of toxins, which may be particularly concentrated by being passed up the food chain, as for example with post-Chernobyl radioactivity, concentrated from the grass consumed by sheep on British hillsides. And, of course, the latest threat has been the possibility of bovine spongiform encephalopathy – BSE, or 'mad cow disease' – 'jumping' species to humans, as it seems to have from sheep to cattle. Concern is increased by the fact that BSE symptoms appear only in its final stages, so infected beasts cannot always be identified.

Medical reports, and the media, customarily highlight a single factor inducing morbidity. Some people take a tougher line, however, disputing the place of any meat, or indeed any animal products at all, in a healthy diet. A woman informant active in the animal rights movement, for example, believes her ethical standpoint to be fully vindicated by the medical evidence:

> 'I just don't understand why people won't wake up to the fact that almost all illness comes from eating animal products. It's just so obvious that vegans are so much more healthy, because they're not continually eating all those dreadful things that are so bad for you. The media will make a fuss about food poisoning, or cholesterol, or anything else as long as it's not looking at the whole picture, which is that we weren't designed to kill and eat other animals. It's like a sort of addictive disease that people have, which they go on doing even when they know it's killing them . . .'

And many believe eating animals to harden our hearts emotionally as well as physiologically. Another set of health concerns associated with meat eating involve mental or spiritual effects – that it may make people cruel, insensitive, or aggressive – and there is some evidence to point to a physical effect. Weinstein and de-Man (1982), for example, tested students on controlled diets and found significantly more negative emotionality amongst meat eaters, results of which they discuss in terms of possible blood chemistry mediation. This link with ferocity can sometimes be put to use, such as in the high meat diet traditionally provided for soldiers. A range of health concerns thus relate not only to concerns about the physical consequences of industrial civilisation's excesses, but also to fears about the spiritual effects our culture's aggressive and exploitative ideology . . . ideas that will be more fully explored in the next chapter. It would be unrealistic entirely to separate the strands of thought, however, as they are inextricably interrelated.

Some scientists have nonetheless continued to cultivate belief in meat's indispensability by casting doubt on the safety of non-conventional diets, caricaturing vegetarians as feebly pallid, in contrast to the (supposedly healthy) ruddy constitutions of their carnivorous counterparts. A distinct genre emerged from the USA in the 1960s and 1970s which is significant for two particular reasons. Firstly individuals and groups concerned are typically treated with barely disguised suspicion as if their subversive beliefs and behaviour threaten more than just conventional nutritional wisdom (which of course they do: in effect they challenge the society's entire cosmology). And secondly it is interesting that such accounts should appear at a time of rapid change in medical orthodoxy, when vegetarian diets were starting to be widely recognised to be at least as healthy as carnivorism, as if some commentators were reluctant to accept the loss of their customary criticism of non-meat diets and sought to defend their own disposition by social stigmatisation instead. The discourse often misrepresents the most extreme practices or short-term dietary treatments as if they were normal behaviour, and is characteristically laced with terminology of 'crazes' and 'faddism' that ideologically marginalises the subjects:

> The effects of faddist diets and cults may be summed up as follows: some fads are harmless in themselves, but many are deleterious and may be damaging to health if used for prolonged periods, e.g. the Zen Macrobiotic diet . . . all strongly indicate a need for more knowledge of the psychological behaviour and emotional appeals that dominate the faddists, and thus an understanding of how to educate them in sound nutritional practices.
>
> (Todhunter 1973: 313)

> Anyone who tries to help these people professionally has to understand each cult in the context of its own terms and must appreciate the influences that govern food-selection if he hopes to effect changes for the better . . . We have found that in making such attempts at education we have to be tolerant and even overlook the abuse that many of these people wish to impose upon themselves. In such cases the best we can do is keep them from abusing the children.
>
> (Erhard 1973: 12)

The continuing production of medical or psychological investigations into such topics as the 'Psychological and cognitive characteristics of vegetarians' (Cooper *et al.* 1985) demonstrates the extent to which some scientists still regard vegetarianism as an aberrant deviation from the norm. Such research says more about the culture which conducts it than about its subjects, not least for assuming that meat eating is the normal behaviour of well-adjusted humans. Their naive acceptance of voluntary explanations such as 'dislike of animal flesh' belies the scientistic sophistication of the authors' terminology, and the entire tenor of the report raises doubts about why the researchers were initially predisposed to suspect culinary minorities of psychopathology. This study employed:

> 8 different psychometric tests including the Hopkins Symptom Checklist (HSC), the illness behavior Questionnaire (IBQ), the Eysenck Personality Inventory, and the Hysteroid-Obsessoid Questionnaire. Results show that although subjects demonstrated elevated somatic concerns on the HSC and the IBQ, they did not differ from control populations cited in the test manuals on other dimensions of psychopathology. Health concerns were indicated as the primary reason for avoiding meat products, followed by the desire to avoid cruelty to animals, dislike of animal flesh, and fear of world food shortage.
>
> (Cooper *et al.* 1985: abstract)

Nowadays, however, overtly hostile, scaremongering reports are increasingly rare, since the bulk of medical evidence has come to show low-meat or non-meat diets to be healthier than a diet centred upon regular intake of large amounts of animal flesh.

It is not only in the late twentieth century that excessive meat consumption has been suggested to be either dispensable or detrimental. Such views have probably been voiced for as long as meat has been eaten. Even a magazine dedicated to body-building at the turn of this century found space to warn in pseudo-scientific terms that red meat should be taken in moderation:

> Don't eat dark meat too exclusively. Some people confine themselves to rare beef and the dark parts of fowl. The dark meat contains a larger amount than other kinds of the irritating stimulant extractive substances of the meat. These are especially unsuitable for nervous individuals and those of a gouty tendency.
>
> (*Health & Strength Magazine* 1901: 358)

Confidence in the nutritional adequacy of vegetable food, and belief in meat's potential unhealthiness have, however, recently gained in circulation. In contrast to the older view that meat is the more easily assimilated, people today are more likely to suggest that 'heavier' meat is more difficult to digest than 'light' vegetables. Red meat especially, consistent with its reputation as the archetypical meat, is now widely felt to be less easy on the stomach than vegetables, or even poultry. Although, to many, meat remains the necessary stuff of healthy nutrition, it has become an alternative modern orthodoxy that animal flesh is a guilty treat to be indulged in only in moderation, if at all. Douglas Adams makes this point in flippant vein by associating hamburger-eating with other unhealthily degenerate indulgences:

> The pimps and hookers, drug-pushers and hamburger salesmen were all outside in the streets and in the hamburger bars. If you wanted quick sex or a dirty fix or, God help you, a hamburger, that was where you went to get it.
>
> (Adams 1988: 195)

Public attitudes to the healthiness of meat have clearly changed. But why now? Progress is conventionally attributed to nutritional and epidemiological scientists' success in gaining a truer picture of the needs of the human body, and to improved dissemination of this knowledge. Colin Tudge suggests there is cause for confidence in modern nutrition, because it addresses all the components of food; it acknowledges the complexity of diet; it is internally consistent; it is compounded of many kinds of evidence; and because 'it looks like good biology' (Tudge 1985: 10–11). Others dismiss such confident advice as mere fashion – after all, it is said, there have always been experts keen to explain their truth, and significant differences of opinion remain.

But whether or not science is really approaching a united and full understanding of our bodily needs, individuals and interest groups continue to derive from it what they will. Just as those who argue that eating animals is ethically wrong can cite medical proof to show it also to be unhealthy – attributing nutritional points in meat's favour to the industry's massive advertising budget, or 'bribery' (Cox 1986: 11) – so the trade finds ample scientific support for meat's nutritional desirability, dismissing fears as faddism.

Nutritional science may have supported changing public attitudes, but cannot fully explain them. For years most people have chosen to indulge their tastes, despite being warned of meat's potential price in terms of health. What is new is not so much knowledge as values. This is reflected in the vigour with which significant numbers of people have begun to examine their diets. George Orwell (1937: 153) wrote that 'the food-crank is by definition a person willing to cut himself off from human society in hopes of adding five years on to the life of his carcase; that is, a person out of touch with common humanity'. Today, however, that crank will find any number of individuals and groups with whom to sociably share her or his concerns.

We must ask why red meat is seen by some as naughty-but-nice, whilst cheddar cheese, for example, with a far higher fat content, is still generally considered good sound nutrition. It may indeed be misleading to talk of saturated fat, or cholesterol: although people may use these terms, it is not the way they think – we are just not that logical:

> The Vegetarian Society launched its Cordon Vert Cookery courses with a press release that featured a dish called Brazilian Bake. The dish derived 77% of its calories from fat; its content of saturated fat per 100g was three times that of lean beef.
>
> (Harrington 1985: 4)

In what way does meat (unlike cheese or nuts) threaten to contaminate? Clues lie in our view of the substance itself. As Mary Douglas has shown (1966), pollution and contagion represent far more than the presence of mere toxins; they signify an ideological threat to *order.* The threat is less to our stomachs than to our thought: to our clear classification of how the world should be. Evidence of the dangers that meat embodies for us can be identified, once again, in the ideas that we hold about it, and about health itself.

Perceptions of what it is to be healthy or unhealthy are the subject of a study by Crawford (1985) which aims to discover how Americans use the word 'health'. He finds a 'consistent and unmistakable theme' running through his interviews:

> Health is discussed in terms of self-control and a set of related concepts that include self-discipline, self-denial, and will power . . . To be healthy is almost equivalent to pursuing health through adopting the appropriate disciplined activities or controls.
>
> (Crawford 1985: 66)

As Crawford explains, the internal logic of the position is not unreasonable. With modern lifestyles that are widely health-denying – sedentary occupations, exposure to harmful substances, and pursual of high-risk activities – discipline is precisely the quality needed to 'negotiate the minefields of health hazards' (1985: 72). However, such a practical analysis still presupposes the pursuit of health to be a rational therapeutic activity, whereas this very perspective 'is integral to an encompassing symbolic order' (Crawford 1985: 73): the reductionist approach to health maintenance is part and parcel of our scientific ethos. Crawford suggests that modern dietary control is better understood as metaphorical, than as literal, health protection.

He presents the image of a situation where a society based on the 'secular religion of capitalism – unlimited growth and continual improvement in living conditions – came up against the objective reality of "limits" – an environment that could no longer support without irreversible damage the weight of unrestrained industrial production', such that the 'social conditions perceived to affect health are too massive, too remote, too unchangeable, [so that] people will normally opt for a course of action within the sphere of personal control'. (People can also take the opposite course and find avenues of *release*, abrogating personal responsibility for health, and finding means of escape from life's troubles, such as eating unhealthy but pleasurable foods (Crawford 1985: 74, 98) – this might shed light on the contemporary popularity of 'junk' food and burger bars.) The pursuit of health, in other words, can be a socially sanctioned outlet for a population who feel threatened by aspects of the world in which they live, and who seek to regain a feeling of security: 'It is possible to say that health is thought about in terms of self-control. It is equally possible to say that self-control is 'thought' through the medium of health' (Crawford 1985: 77).

Diet is a focus for control since the danger to our wellbeing increasingly is seen to come not only from isolated sources of contamination, solvable by further technological fixes, but from our entire life support system: the polluted planet. For example, when large numbers of seals dying in the North Sea came to public attention in 1988, it was evident that despite the efforts of media experts to reduce the problem to a viral epidemic, there was a widespread view amongst the general public that the disease was better understood as a symptom of the wider problem of excessive sea pollution, with clear implications for human life. In spite of our efforts to avert our gaze from intimations of animality, we remain dependent on the natural environment for the foods that sustain us, and vulnerable to corruption of the nourishment which we derive directly from our habitat.

Meat is inevitably affected by these changing attitudes. Even the Director of Planning and Development of the British Meat & Livestock Commission acknowledges that 'dietary advice and associated health issues will inevitably act as a depressant on the demand for red meat and meat products' since, he says, visible fat on meat reinforces consumers' 'other objections' to the product, particularly as 'promulgated by interest groups'; since 'it is easy to create concern about what is used as a raw material for the product'; and since 'there is a backlash against technology which manipulates the raw material to change its form and composition' (Harrington 1985: 23)

It is interesting that Harrington should refer to so many associated issues. For health concerns more than the appraisal of material threats to the physical body, with subsequent application of sophisticated therapy by clinical experts, that allopathic medicine characteristically implies. Such 'conventional' medical approaches seek, as Ivan Illich puts it, 'to engineer the dreams of reason' (1976: 47). But health is simpler than that. It is, in the end, 'simply an everyday word that is used to designate the intensity with which individuals cope with their internal states and their environmental conditions' (Illich 1976: 14). Health is about *how well people feel,* and a person's entire set of beliefs and feelings is relevant to their health.

Illich argues that today's costly medical bureaucracies are in fact health-denying – 'not in their instrumental but in their symbolic function: they all stress delivery of repair and maintenance of the human component of the megamachine' (1976: 69), focusing upon isolated causes of illness, and ignoring the many additional determinants of good or poor health. They usurp the individual's responsibility and make it the preserve of the specialist. By the imposition of a model of technological medicine, the individual is shaped into the model of modern industrial culture, as medicine in every culture performs a similar function (Zola 1972).

Decisions by members of our society to alter their eating habits may thus be more significant than is immediately obvious. The trend, by some at least, towards eating either less meat, or less of the meats perceived to be unhealthy (such as red meat), as part of a search for control in a threatening environment, can be interpreted in terms of a movement towards regaining personal responsibility for health, and thus against institutional medical care. This must be seen against a background of rapid growth in the market for healthier food in general (including consumer resistance to chemical additives, and demand for wholefoods and organic produce), expanding interest in exercise, and the increasing popularity of so-called holistic approaches to medical practice in which individuals are treated not as mechanical devices with faulty parts, but rather as whole beings in a dynamic relationship with the environment upon which they depend. In this context, the evidence of a widespread reaction against the industrialisation of everyday life is strong, and changing attitudes to meat are one part of that process.

A consistent thread is evident in the many and various health concerns associated with meat and other foods. They suggest rejection not only of modern society's unhealthy nutritional habits, but also of the implicit non-naturalness of many of the values of our civilisation. A good example is the notorious television 'sausage programme' (*World in Action*, 7 October 1985) which gave British consumers a vivid insight into the production of Mechanically Recovered Meat, after which sales of sausages and other processed meat products fell dramatically, never fully to recover. Above all, it was the artificiality of the mechanical process whereby heavy-duty industrial equipment is fed with parts of animals that most chefs would regard as beneath consideration, to be processed and disgorged as a pink slurry for concoction into human foodstuffs such as pies and patés, that offended viewer's sensibilities.

Or consider the various 'health scares' that intermittently preoccupy the media. At first sight the issues are diverse and unrelated: listeria; salmonella; irradiation; BST; BSE; cook/chill; additives; hormones; antibiotics; pesticides; unhygienic slaughterhouses . . . the list seems endless. Yet they have a common factor: in each case the threat is perceived to originate in some aspect of 'unnatural', or technological, food supply. This is not to say that there is no physical basis for the concerns; technological interference in natural processes has indeed bequeathed us with a seemingly endless repertoire of previously unimagined threats. These are the issues in which our reductionist society, that demands isolated causal relationships for scientific and political credibility, conducts its discourse, but seen together they express more. Just as the consumption of convenience food symbolises participation in modern consumer society, so such health fears metaphorically signal public loss of confidence in the excesses of a food production and supply system that, by and large, had hitherto been trusted.

Health care has accordingly paid increasing attention to the environmental origins of our countless endemic ailments. We are told, for example, that most of us consume too much fat. Fatness in people was once thought (like meat) to indicate prosperity, but today is regarded – by the western middle classes at least – as a sign of sloth, indiscipline, and decadence. And since nature is thought to be a state where the fittest survive, so fat has come to signify the unfitness of unnatural culture.

The suggested link between meat eating and raised blood pressure (another hallmark of modern civilisation) also relates meat and ill health to industrial culture. A series of studies in the medical journals in recent years has indicated that those on vegetarian diets suffer less from rising blood pressure with age than the general population, that adoption of such a diet could result in a fall in systolic blood pressure (Sacks *et al.* 1975; Armstrong *et al.* 1977; Rouse *et al.* 1983a, 1983b; Margetts *et al.* 1986). It is significant that each of these studies specifically deals with various vegetarian diets rather than, say, high-fibre or low saturated fat diets.

Most doctors will give dietary advice in response to certain complaints. In particular, attention to food intake is seen as a necessary and effective response to complaints including arthritis and cancer, gastric ulcers, diabetes, and heart disease – in other words to the afflictions of the modern, affluent, industrial world. But in much complementary health care diet is not just part of treatment but is a central consideration, since food is our major interaction with the environment that sustains or contaminates us. And, as one naturopathic practitioner advises:

> 'If there's one single piece of advice I'd give to my patients – apart from maybe not to jump under buses – it's to eat less meat, and far more fresh fruit and vegetables. It's amazing to see the results when people do that: within a few days or weeks they find that the rheumatism or depression or migraines or whatever that's bothered them for years just stop happening!'

When illness is seen to be symptomatic of deeply rooted industrial contamination, as is the case with many complementary therapies, meat is the foodstuff that is most often singled out for mention. It stands indicted as the worst example of industrial culture gone astray. Many of the assumptions upon which we are reared are confronted and refuted, criticised for being predicated on the wrong values. For example, whereas meat, or 'protein', is conventionally extolled for helping people to grow strong, the authors of a bestselling book which advocates a diet consisting almost exclusively of raw vegetable foods as the way to full health also deny that meat, as a high protein food, should have any place at all:

> So insidious and destructive are the effects of a high protein diet, and so extensive is the research which proves as much, that is is difficult to understand why the 'lots of protein is good for you' myth still survives. Excess protein is so damningly implicated in premature aging that it is hard to understand how anyone who is serious about caring for themselves in the long term can continue to eat large quantities of high protein foods . . . Lots of protein certainly brings about early and rapid growth, but it also brings about early and rapid ageing and disease.
>
> (Kenton and Kenton 1984: 94)

And a text that takes a radical approach to the AIDS controversy – which argues that the syndrome is essentially a symptom of general poor health due to environmental pollution, devitalised foods, drug abuse (not least by the medical profession and farming industry), and loss of spiritual direction – suggests that a holistic path to regaining health should be a self-treatment programme including careful attention to food intake, and in this case it is not just meat in general that is proscribed – here the association with industrially reared livestock is explicit:

> Meat which is produced under modern 'factory farm' conditions is thoroughly undesirable for a number of reasons, including the high saturated fat content and possible hormonal and antibiotic residues. The only flesh foods eaten should come from free living animals . . .
>
> (Chaitow and Martin 1988: 184)

But the particular causative connection advanced for each illness associated with meat, accurate or inaccurate, is perhaps of less significance than the general observation that it is industrial civilisation, through the medium of chemicals or other residues, or through the misapplication of unnatural processes, that is indicted. Fears of contamination are by no means restricted to meat but, partly because meat comes from animals high on the food chain and so may contain a higher concentration of toxins than an equivalent weight of vegetable matter, partly because excessive consumption of animal fat or of protein (like excessive consumption of almost anything) may indeed be harmful, and partly because animal flesh as a food epitomises for us an exploitative relationship with nature, it is an ideal channel through which to express such concerns. This points towards a deep-rooted change not only in attitudes to meat, but in society's view of the world in general, and in particular of humanity's relationship with that environment.

Many today seek the 'natural' human diet. Whilst no single such diet can ever have existed, the quest for a pattern of eating more closely related to that upon which we evolved has clearly developed in response to the highly processed products of the modern food industry. It is part of a more general trend whereby society's optimism about the infinite possibilities for a roseate future under absolute human control, expressed most vibrantly in nineteenth- and twentieth-century science fiction, has been replaced by widespread apocalyptic fear for the future, distrust of the present, and nostalgic hankering for a cleaner, healthier, better, bygone age. Utopia has suddenly migrated from time future to time past, without our ever having been aware of experiencing it in time present.

The meat trade is aware of its vulnerability on this issue, knowing that customers prefer at least to imagine their food to be as wholesome as that of olden days. A common response is to invest meat products with rustic associations by expensive image-building, tapping folk memory for scenes that bear little relation to the intensively regulated 'factory-floor' conditions in which most modern livestock is confined between birth and slaughter. Meat products, like other industrially processed foods, have been emblazoned with claims to natural goodness, so that by June 1989 the British government was moved to publish a code of practice for the use of terms such as 'natural'. It remains to be seen, in any case, whether such marketing techniques can be successful whilst meat is so intractably associated with the conquest of nature.

In Chapter 4 it was shown that food is a ubiquitous metaphor through which we demarcate our cultural identity, and by which we characterise feared or despised outsiders. The unique aspect of our situation today is that it is apparently our very own attitudes that are distrusted and seen as suspect, our very own technological successes that are now perceived as a threat. Today it seems that we ourselves are the enemy.

13

ETHICS

> 'I knew I shouldn't eat it for my health but I didn't listen to that, but then I thought what they did to the animals . . . Yes.'

Animal Rights has become something of a *cause célèbre* in recent years, partly due to the publicity given to the activities of the more zealous protagonists. Attacks on the homes of suspected vivisectionists, the fire-bombing and window-breaking of stores selling animal furs, attempts to contaminate hamburger supplies, and a range of other illegal actions, although publicly condemned, reflect the intensity with which some believe that society is engaged in atrocities against non-human animals. In the United States, members of animal liberation and environmental groups stage sabotage raids on ranches, seeking 'the elimination of the livestock industry' (Associated Press 1989: B7), and 'sabbing', as members of the British hunt sabotage movement refer to their actions, can today attract as many followers as the hunts against which they set themselves. Even the consumption of any meat at all can arouse strong feelings amongst abstainers, not least recent converts:

> 'At first I was really fanatical. I used to go around everyone when they were eating and saying "You shouldn't eat meat; it's wrong", and at my parents at home I would take away all the meat out of the fridge and throw it over the wall in the yard. They got a bit fed up because it was proving quite expensive. But I've calmed down now.'

Even a British government minister was recently touched by the public mood of concern for animal welfare, when he sought to require fur garment labels to state if it was likely that animals used had been caught in a type of leg trap seen as particularly cruel (Vulliamy 1988: 6). Although the legislation was ultimately dropped, allegedly under Canadian threats to cancel an order for a British nuclear installation, the very appearance of such proposals marks the increasingly broad base throughout society that arguments, which would until recently have been regarded as unreasonably extreme, have today won. Animal testing of cosmetics has similarly fallen out of favour, as evidenced by the commercial success in the 1980s of Anita Roddick's Body Shop, and its influence in catalysing more conservative firms to follow suit.

Mainstream opinion has moved from the assumption that animals have no rights whatsoever, their interests being entirely subject to those of humans, to growing agreement that they deserve some consideration, with some believing that they have inalienable rights. Legislation governing the scope and nature of experimentation on live animals has been incrementally introduced, although its constraints are still far from sufficiently stringent to satisfy the more vigorous critics. The liberal pole of opinion in the new animal rights movement is committed to goals including abolition of the use of animals in science, dissolution of commercial animal agriculture, and elimination of commercial and sport hunting and trapping (Regan 1985: 13). In other words it is the *extent* of animals' rights that is now the matter in question, rather than their existence.

To illustrate: the biblical phrase 'do unto others' presents a basic ethical dilemma: namely how far 'others' extends beyond the self. Different peoples have drawn the limits of consideration at widely various thresholds, but their extension to animals is by no means new. Leonardo da Vinci was teased by his friends for being so concerned about the sufferings of animals that he became vegetarian (McCurdy 1932: 78), and Plutarch warned: 'Let us eat flesh, but only for hunger not for wantonness. Let us kill an animal; but let us do it with sorrow and pity and not abusing it or tormenting it, as many today are wont to do' (quoted in Pullar 1970: 226). In ancient Rome (as in contemporary Spain), however, animals' suffering made popular entertainment:

> At one time a bear and a bull, chained together, rolled in fierce combat across the sand; at another, criminals dressed in the skins of wild beasts were thrown to bulls, which were maddened by red-hot irons, or by darts tipped with burning

> pitch. Four hundred bears were killed on a single day under Caligula . . . Under Nero, four hundred tigers fought with bulls and elephants . . . In a single day, at the dedication of the Colosseum by Titus, five thousand animals perished. Under Trajan, the games continued for one hundred and twenty-three successive days. Lions, tigers, elephants, rhinoceroses, hippopotami, giraffes, bulls, stags, even crocodiles and serpents were employed to give novelty to the spectacle.
>
> (Lecky 1869: 118)

This does not prove that the Romans were entirely without morality. As Singer notes, they 'showed a high regard for justice, public duty, and even kindness to others. What the games show, with hideous clarity, is that there was a sharp limit to these moral feelings'. It would have been outrageous to have treated a respected human in such a manner, but for those outside that realm, including some criminals, prisoners of war, and all animals, there were no such qualms (Singer 1976: 208).

Keith Thomas (1983) abundantly illustrates the seventeenth- and eighteenth-century heritage of modern relations between Man and the Natural World. Certainly, the early modern period saw a new conviction in the elevation that philosophers, scientists, and theologians accorded to the human species. In catechistical doctrine, according to Lancelot Andrewes, animals had no rights: they 'can have no right of society with us, because they want reason' (1650: 217); Bishop Ezekiel Hopkins declared that 'We may put them to any kind of death that the necessity either of our food or physic will require' (1692: ii, 3); Wollaston also opined that the sufferings of brutes are not like the sufferings of men, as they have no conception of the future and lose nothing by being deprived of life (1722: 34–35); and the nonconformist Philip Doddridge believed that because animals are 'capable of but small degrees of happiness in comparison with man [it is] fit that their interests should give way to that of the human species whenever in any considerable article they come in competition with each other', pointing out that the instinct that brings fish in shoals to the sea-shore 'seems an intimation that they are intended for human use' (1763: 130, 133).

Descartes is widely credited with popularising the notion that non-human animals are mere automata, incapable of suffering, lacking the soul that distinguishes humans, and thus undeserving of 'humane' consideration. His message was seized upon by a scientific, technological, and industrial establishment, eager to investigate and exploit the living world. Descartes distilled the spirit of an age when industry and technology were coming rapidly to the fore.

But even at this time not everyone was of one accord, and with the Enlightenment in particular more voices came to be raised in animals' defence. David Hume talked of 'gentle usage' to animals; Rousseau's idea of the 'noble savage' brought a new appreciation of nature, if a somewhat romantic one; Alexander Pope argued that we were also responsible to God for the 'mismanagement' of animals; and Voltaire compared Christian practices unfavourably with Hindu (Singer 1976: 220–223):

> Barbarians seize this dog who so prodigiously surpasses man in friendship. They nail him to a table and dissect him alive to show you the mesenteric veins. You discover in him all the same organs of feeling that you possess. Answer me, mechanist, has nature arranged all the springs of feeling in this animal in order that he should not feel?
>
> (Voltaire 1764: 65)

But perhaps the most famous contribution of the period to the 'humane movement' is that of Jeremy Bentham:

> The day may come when the rest of the animal creation may acquire those rights which never could have been withholden from them but by the hand of tyranny. The French have already discovered that the blackness of the skin is no reason why a human being should be abandoned without redress to the caprice of a tormentor. It may one day come to be recognized that the number of legs, the villocity of the skin, or the termination of the *os sacrum* are reasons equally insufficient for abandoning a sensitive being to the same fate. What else is it that should trace the insuperable line? Is it the faculty of reason, or perhaps the faculty of discourse? But a full-grown horse or dog is beyond comparison a more rational, as well as a more conversable animal, than an infant of a day or a week or even a month old. But suppose they were otherwise, what would it avail? The question is not, Can they *reason*? nor Can they *talk*? but, Can they *suffer*?
>
> (Bentham 1789: Ch.17)

Nonetheless, Bentham, in common with most moralists of the period, was concerned only with 'unnecessary' suffering, and did not himself abstain from meat. Singer argues that the nineteenth-century anti-cruelty movement rested on the assumption that the interests of non-human animals deserve protection only when serious human interests are not at stake – endorsing the official ideology of recent western society which places the premier ethical boundary around *Homo sapiens.*

The significance of the new animal liberation movement is its challenge to this assumption. Taken in itself, say the animal liberationists, membership of the human species is not morally relevant (Singer 1976: 4). The modern movement seeks to push the perimeter of the ethical net yet wider, arguing, for example, that if 'possessing a higher degree of intelligence does not entitle one human to use another for his own needs, how can it entitle humans to exploit non-humans for the same purpose?' (Singer 1976: 7). It seeks to bring other animals into the fold of humanity.

Striking comparisons can be (and are) drawn with other liberation movements in history, such as women's emancipation and the abolition of slavery. Singer, for example, entitles his first chapter of *Animal Liberation* 'All Animals are Equal . . . or why supporters of liberation for Blacks and Women should support Animal Liberation too' (1976: 1), and Carol Adams writes that 'being a vegetarian reverberates with feminist meaning' (1990: 13). Throughout history individuals have discerned unjust abuse by powerful 'in' groups of those falling outwith the obtaining realm of consideration, disputing what is seen as an arbitrary and inequitable division between the users and the used. Thus, today, to discriminate between humans on grounds of faith, gender, or skin colour is widely, if not universally, agreed to be unacceptable, whilst for those on the margins – or perhaps in the vanguard – of the ethical debate there may be a commonality of interest in opposing apparent exploitation in whatever form it is found:

> 'I don't want to eat anything that's suffered for my benefit. I can't really enjoy food that's suffered. The thing that made me aware in the first place was factory farming, but since then I think I've become aware like most vegetarians that there's a lot of different reasons. There's the health aspects, though that isn't one of my main reasons; and the unwillingness to kill animals, or have them killed for my benefit; and also, if

> we weren't as a country feeding tons of grain to our animals to produce protein, then I think there'd be a lot more food to go around the rest of the world.'

> 'And also, the whole thing about being vegetarian has now also become very associated with other things. There are sort of three things that go together, which are the bicycle, being a member of CND, and being vegetarian. So if you're one, you have to be the other two.'

In fact vegetarians exist in every walk of life who do not fit the above stereotype, but its existence perhaps reflects an element of truth. Vegetarianism tends to be linked with a range of 'progressive' concerns, as an integral part of a personal set of linked beliefs, although the particular concerns will of course vary from person to person. An interesting example is the woman who says that she would prefer not to have to buy leather shoes, but does so on the grounds that anything else is bad for her feet. She reconciles this moral dilemma, she says, by buying hand-made shoes from a workers' co-operative, which she sees as a sort of moral counterweight:

> 'I would call myself a conditional vegetarian, by which I mean I don't mind eating meat sometimes as long as I know that the animal has been well looked after, because I think that's fairly natural.'

There is widespread public sensitivity to the prevailing attitude within much of the food and farming community which regards care for their animal raw materials as little more than a commercial oncost. This has stimulated the development of a considerable market for free-range and organic meats and animal products, deriving from more humanely reared livestock. Many people are willing to pay substantially more for such items.

Meat suppliers do not of course willingly concede the moral high-ground, maintaining the consumption of animal flesh to be a perfectly natural or necessary aspect of human life. And, if directly questioned, most people would probably advance some justification for the rearing of animals for food:

> 'I always go back to the fact of our having carnivore teeth. That shows that we're meant to live on meat surely, doesn't it.'

'It's a natural part of the food chain. That's what lions do in Africa to live.'

'What protein alternatives would you say there are to have?'

Food industry orthodoxy still regards its animal raw materials as little more than meat machines, supplying a product demanded by consumers. The meat business is regarded as a business like any other, whose merchandise happens to be made from animals:

> 'The company before I joined them was having all sorts of management problems . . . And at that time I was the sales director of [a] sausage manufacturers, so they thought that I could make the fairly quick transition, given that I was sort of interested already. In manufacturing, that is. And that was beef as well – it was to do with creating the raw material from which sausages are made.'

Within farming circles the traditional 'argument in favour of [intensive] systems is that the stock have to be content to be healthy and healthy to be productive; thus the farmer has a strong vested interest in ensuring they are content'; this is being modified in recognition that 'many people now consider the price of progress is too high and that the animals' needs have to be examined more carefully' (Robertson 1989a: 18). This apparent conversion towards the principle of a measure of care for the feelings of animals seems to indicate some change from the attitude of a 1929 trade textbook whose sole concession to avoiding the condemned animals' suffering was still for human profit. It recommended that 'slaughter immediately upon arrival [at the abattoir] is universally condemned' because exhausted animals bleed imperfectly, and that blows of all kinds should be avoided as they can cause 'unsightly appearance in the meat' (Hammet and Nevell 1929: 134). But the sincerity of the industry's conversion to the cause is widely doubted since welfare remains primarily a marketing issue:

> 'Well, of course they'll say that they're treating animals well, but who believes that? Just look at all the conditions chickens are kept in . . . even so-called free range ones! No, they're only in it for another fast buck. In the end their only interest is to kill the poor animals to sell to be eaten. That's caring?'

Such scepticism could only be reinforced by reading the trade press. The *Meat Trades Journal* (1989g), for example, bears the headline: 'Good welfare saves money', reporting that 'Abattoirs and farmers are losing £5m a year through damage to animals in transit to slaughterhouses, according to animal welfare experts', since animals suffering from stress, bruising, and marks fetch a lower price. Discussion of animals' welfare for their own sake remains rare.

But the once-conventional view that meat is an entirely necessary and natural part of the diet is now contentious. Even for those with no overt objection to the farming of animals, the industry recognises that their raw material is emotive, and that its promotion must be undertaken with care. Accordingly, direct reference to ethical questions pertaining to meat tends not to be pursued. Instead a more subtle approach is taken to cultivating a positive ethical image, such as the 1987 promotion at Gloucestershire schools, where for every red meat meal bought by children at lunch-times British Meat contributed 10 pence towards buying and training a guide dog. The promotion of red meat (symbolising power and control) as aiding the blind (a relatively weak and disadvantaged group) is intriguing.

The meat industry is consistent in emphasising the importance of supposedly rational influences such as health and economics, rather than emotive ethical arguments. This can partly be accounted for by a deliberate policy of avoiding issues on which the trade might be vulnerable but seems also to be partly due to a genuine lack of understanding of the concern in many people's thought and consequently as a determinant of their behaviour. To the meat industry executive, used to thinking of livestock only as productive units, for example, the extent of such alternative conceptions may be difficult to comprehend:

> *Q. 'What about the more moral complaints that some people have about meat?'*
>
> 'I don't know if that's anything like as prodigious as the health aspect . . . I think that when you look at the morality of it you always get a thin spectrum of the community who feel sensitively, and I'm not in any way denigrating them, who feel justified in protesting that "we don't wear fur coats and we don't kill animals actually for food" . . . And there's nothing wrong with that. We're democratic, and we feel

> that a certain section of the population ought to be allowed to express themselves in that way as long as they keep it within the limits of the law. But I wouldn't say that that would have a major influence. Nothing like the influence of people listening to the medical evidence . . . Although it gets publicity which is probably disproportional. It hangs there, as a sort of afterthought.'

However, when a representative of the British Meat & Livestock Commission says that . . .

> Recent reports from the Vegetarian Society show that it recognises that debate of the ethical issues will not achieve many converts and that it intends to concentrate on welfare, conservation and health issues . . .
>
> (Harrington 1985: 4)

. . . it is not obvious what is meant by ethical concerns, if such issues are excluded. Ethics, welfare, conservation, and even health, are indivisible, and have indeed been significantly influential in changing attitudes to meat, although the process is more subtle than a matter of individuals eschewing meat on coming to rational moral decisions concerning cruelty. To understand the meat system in its ethical context, it is useful to consider the issues on a spectrum, rather than each particular cause in isolation. In this way we can identify a range of attitudes, from the most narrowly conservative, who might argue that we have social responsibilities only for ourselves as individuals and for our families, and for whom unbridled exploitation of any being outwith that sphere is legitimate activity, to the broadest and most liberal, who might perceive a duty of care for their entire environment, living and inanimate.

The meat industry, in common with most commercial business, falls squarely into the former camp. Its characteristic attitude is that morality, as such, is the concern only of the individual consumer. ('Yes, as a sheep producer, we're only supplying what the public wants.') Its legitimate interest is in profitable production; not in the ethics of that production. Abrogation of such responsibility must, however, be seen as somewhat disingenuous when the consumer – whose conscience is allocated the task of decision, and who is already divorced from participation in the process – is systematically shielded from the reality of modern breeding, farming methods, and slaughter, by secrecy and by the sophisticated marketing of a mythical past.

Religions are often regarded as the ultimate arbiters of ethical behaviour, at least by their adherents, so it is interesting to consider their views of the correct treatment of animals, and of meat. And in the Judeo-Christian tradition in particular, meat has again broadly represented human power. Peter Singer, for example, suggests that in the Bible there is little doubt about our proper relationship with the mortal world:

> After the Fall of man (for which the Bible holds a woman and an animal responsible), killing animals clearly was permissible. God clothed Adam and Eve in animal skins before driving them out of the Garden of Eden. Their son Abel was a keeper of sheep and made offerings of his flock to the Lord. Then came the flood, when the rest of creation was nearly wiped out to punish man for his wickedness. When the waters subsided Noah thanked God by making burnt offerings 'of every clean beast, and of every clean fowl'. In return, God blessed Noah, and gave the final seal to man's dominion.
>
> (Singer 1976: 204–205)

In curious resonance with the profane assumption, noted in Chapter 5, that the origin of the human species can be dated to the point at which we began to hunt and kill other animals, many theologians and lay Christians ascribe the beginning of meat eating to the Fall, although arguments have raged as to whether it was the human constitution or vegetable food (on which we lived in Eden) which had degenerated, or whether agricultural labour required more robust sustenance (Thomas 1983: 289). In Hebrew lore it is Noah who is said to have been the first non-vegetarian (Chiltosky 1975: 235–244).

And when Abel, son of Adam and Eve, sacrificed his beasts to the Lord, or when Noah made burnt offerings, they used animals. For if human power over the mortal world is represented by cooking and eating meat, then destruction of animals by the extreme mediation of fire can be interpreted as unequivocal acknowledgement of God's greater might. It states that our worldly power ultimately derives from the higher being's unworldly power, to which homage must be paid. Bestowed by the ultimate blessing – divine proclamation – meat has operated as a symbol of the demigod-like status of humanity, since God is said to have given us dominion over every living thing. Peter Singer comments that when it is said that God created Man in His own image, we might fairly regard this as man creating God in his (1976: 204). In structural terms, we thus position ourselves relative to the rest of creation as we believe God is to us: omnipotent. (Science and technology have meanwhile endeavoured to emulate God's omniscience and omnipresence, by which our worldly divinity might be complete.)

At Christian communion material is physically consumed that is believed, by transubstantiation, to have become the body of Christ. By an inverted logic that doubly negates normal values, we thus cannibalistically consume a greater being, and homeopathically absorb some of His spiritual strength. This can be interpreted as obliquely serving to sanction our mundane consumption of lesser beings. Meat remains the archetypical representation of control, but in these contexts alone its role is to demonstrate that our power is limited: we are masters only of all we survey; the supernatural is beyond our grasp.

Meat is the food most strongly proscribed on fast-days: days when the greater glory of God is especially to be honoured. Thus Trappist monks have long abstained from meat, considering it a luxury incompatible with their vows of simplicity (Majumder 1972: 175). This goes back at least to medieval days, when rejection of meat by the devout occurred in a context of denial of the flesh that drew directly on manichaean conceptions of bodily affairs as 'totally evil, all nature as corruption, and the cessation of physical being as the proper end' (Twigg 1983: 19). In this predominantly negative concept, Twigg argues, there was little sense of vegetarian food being 'higher' food, as is common today. However, by its reputation as *less* defiling, it does indicate a view of meat as *more* associated with love of profanity. The church has a long history of commending its avoidance when spiritual control is held to be the particular ambition of good Christians.

There is little disagreement that, over the years, church dogma has usually condoned the prevailing use of animals as rightful expression of the natural hierarchy. One of the most influential treatments of this subject in recent years has been that of Lynn White Jnr. (1967), who suggests that Christian theology has legitimated a fundamentally exploitative relationship between humans and the rest of the natural world, and is thus a root cause of the modern ecological crisis. Many Christians today are indeed troubled by the problem of reconciling their religion's cosmology, whose roots lie in the ancient Hebrew world, with twentieth-century experience. The interpretation and application of the church's teaching on human dominion, as it is still widely perceived, may seem to be at odds with its basic message of compassion:

> 'Because I know that – as I said earlier, obviously I'm a Christian – and I know we were given anything virtually to enjoy . . . But, I don't think God ever imagined that we were going to treat creatures in the way that we do . . . So you sit on your hands and keep quiet . . . And as long as you're patient and love people, then, eventually, God has His way. But, it's quite stressful in some ways.'

Nowhere in the Gospels is there any explicit statement recorded on the part of Jesus that cruelty to animals is one of the great sins (Niven 1967: 23), and the Bible explicitly tells us that God created Man in His own image to have dominion over 'every living thing that moveth upon the earth' (Genesis 1: 28). This idea, more than any other, writes Andrew Linzey, 'has characterised much Christian thinking about our treatment of animals. The standard historical interpretation of this verse is that we may use animals for our own betterment and happiness (1985: 11). *A Catholic Dictionary,* for example, describes animals as 'not created by God, but . . . derived with their bodies from their parents by natural generation' (Addis and Arnold 1924: 31), and Roman Catholic commentaries still tend to hold that 'We have no duties of justice or charity towards them [but only] duties concerning them and the right use we make of them' (Davis 1946: 258). 'But this interpretation,' Linzey argues, 'vastly influential as it has been in the past, finds little theological support today' (1985: 11).

Linzey sets out to demonstrate that the Gospels are not in fact as anthropocentric as they have long been interpreted to be. 'If this is right', he says, 'it means nothing less than for centuries Christians have misinterpreted their own scripture and have read into it implications that simply were not there' (1985: 10). He points out, for example, that in verses 29 and 30 of Genesis 1, humans are 'commanded to eat "every plant yielding seed" and "every tree with seed in its fruit" for food, that is, to be vegetarian . . . It is hardly likely if the concept of dominion meant absolute power over animals that there should be a divine prohibition concerning the eating of them' (Linzey 1985: 11–12).

It is significant that modern commentators can find biblical justification for views which contradict traditional dogma regarding humankind's unrivalled supremacy in, and thus freedom to exploit, the natural order. In fact, whether or not the Bible is itself intrinsically anthropocentric is relatively unimportant, whilst it is incontestable that exploitation rather than stewardship has consistently been the dominant theme (Thomas 1983: 24–25). But much as science regularly provides the answers that its lay and specialist public desires, so too religious teaching is anything but free of its cultural context. Like their forebears, modern theologians do not only guide, but also follow – and are having to revise their interpretation of the Bible's doctrine on the subject of environmental ethics because the subject is of deep and widespread concern.

Other spiritual teachings share many elements in common with Christianity in their attitudes to meat, but many differ markedly in their traditional view of the proper human attitude to the natural world. 'Concern for animal suffering can be found in Hindu thought, and the Buddhist idea of compassion is a universal one, extending to animals as well as humans' (Singer 1976: 2). These religions have long counselled to a greater or lesser degree against the eating of meat, even where there is no shortage of animal foods (Dwyer *et al.* 1973).

Eastern teachings typically regard earthly life as limited or illusory. From this point of view, meat is a temptation to be avoided, not only because cruelty is considered to engender insensitivity, but also because the enjoyment of meat is a sign of attachment to mere worldly power, whereas spiritual strength is the true objective to be pursued. In India, for example, vegetarianism is recognised by the whole of the Hindu population as the superior form of diet and as a reflection of high civilisation; it is therefore particularly practised by Brahmans as a sign of social and spiritual stature (Dumont 1972: 190–195). Partial or total abstinence from meat, as a sign of voluntary simplicity or as an act of self-discipline, is an almost mandatory aspect of prescribed paths to enlightenment ranging from Hatha

Yoga to Seventh Day Adventism (Bernard 1982: 85n; Todhunter 1973).

Norms do not of course always match practice. Whilst it is a Buddhist principle, for example, that one 'may not knowingly deprive any creature of life, not even a worm or an ant' (Westermarck 1924: 497), 'In many Buddhist areas, including Tibet, Ceylon, Burma and Thailand, even Buddhist priests eat meat' (Simoons 1967: 10). In parts of India such as eastern Uttar Pradesh even Brahmans will eat meat (Dumont 1972: 184). In 'the Middle East and the Mediterranean area too, vegetarian practices have generally been observed scrupulously only by clergy and the very devout. Vegetarianism was common in ancient Persia only in the priestly and learned class of the Magi' (Simoons 1967: 11). The Mandaeans of Iraq and Iran are also reported to regard animal slaughter as immoral and degrading – and so to be apologetic for their sinfulness when doing so (Drowser 1937: 48, 50). But however scrupulously proscriptions are observed by fallible mortals, the contrast between the conventional attitude to meat in such belief systems, and in orthodox Christianity, is striking.

Whilst Christian ethics have long traversed the world under the patronage of western commercial and religious imperialism, Eastern spiritual influences have increasingly pervaded the west in recent years. Buddhism, Hinduism, and other derivative sects, have had a significant effect in changing, or at least catalysing change in, many people's views of meat. Novelist Richard Bach, for example, influenced by such ideas, associates meat consumption with a brutal view of the world – typical of a society with which he no longer feels fully at one – whereas his learning to perceive the possibility in a more spiritual life brings with it respect for other beings:

> Once I would have ordered bacon or sausage for this meal, but not lately. The more I had come to believe in the indestructibility of life, the less I wanted to be a part of even illusory killings. If one pig in a million might have a chance for a contemplative lifetime instead of being skrockled up for my breakfast, it was worth swearing off meat. Hot lemon pie, any day.
>
> (Bach 1985: 23–24)

'I'd felt uneasy about meat for a long time, I suppose, but I'd never let it really bother me too much – until I moved into a house sharing with a couple who were both Buddhists . . . it did make me realise that the way we treat animals – and especially eating meat – isn't just cruel to them, but it has a horrible effect on us ourselves too. And I think I just wanted to stop, once I'd sort of realised that.'

Q. 'What does Hare Krisna teach then?'

'It's because they're trying to develop a peaceful state, and if you're eating something that's going to involve suffering and bloodshed then . . . it's like what you eat. It affects your consciousness, and it's just completely unnecessary to cause violence towards animals.'

Similar ideas have found an outlet in western society through such avenues as the growing popularity of meditative practices, which typically place considerable emphasis on the need for the participant to improve their entire way of life, including diet, to achieve the desired progress. The popularity of such teachings, particularly amongst the young, can be seen as meeting a spiritual need in a society disillusioned with its own traditions. With the consensus of public feeling seemingly shifting in favour of extending a degree of ethical consideration further than the human species, such teachings offer an alternative to the revisionism that has occurred within Christianity, as a spiritual basis for a more enlightened attitude to our dealings with ourselves, and with the planet upon which we live.

14

ECOLOGY

The litany of social and environmental sins associated with meat production could almost make one feel that meat alone is responsible for bringing the world to a state of ecological crisis. The dangers may be real and the connections by which meat is blamed for worsening them genuine but meat is so widely indicted also because its consumption is an apt *symbol* for incrimination.

What meat exemplifies, more than anything, is an attitude: the masculine world view that ubiquitously perceives, values, and legitimates hierarchical domination of nature, of women, and of other men and, as its corollary, devalues less domineering modes of interaction between humans and with the rest of nature. It is this entrenched attitude which, now taken to its logical and perhaps even terminal conclusion, has brought the very habitat on which we depend to its current perilous condition. And it is inescapable evidence of the contingent apocalypse having already arrived for many human communities and ecosystems, and of impending catastrophe almost everywhere, which has led at last to a questioning of the values on which we found our cosmology. The environmental control that meat still represents has taken on negative implications for many people today, superseding the predominantly positive meanings of the past.

But first, let us consider a few of the environmental issues in which high meat consumption is involved. Perhaps the most common negative perception of meat in terms of the ecology of the planet is its association with poverty and starvation amongst the world's politically and economically disadvantaged. At its simplest, the sight of the affluent west with food literally to waste, and luxurious meat in plenty, whilst people elsewhere die of starvation, can arouse simple feelings of injustice:

> 'And it's the whole guilt thing about people on the other side of the world not having enough food to get them through the day; it's the whole . . . Food has become a terrible . . . guilt . . . problem. Especially nowadays when there's actually a surplus of meat. It'd be okay if they could get things sussed out so they were actually going to feed the people that are needing it, but that's a quantum leap. There's no government that's ever going to.'

Misgivings may be strengthened by awareness of the low energy efficiency of meat production – the conversion of grain into animal flesh requires about ten calories for every calorie provided for human consumption, or five grams of protein input for the production of one gram of meat protein; for beef the ratio is more like twenty to one (Pimental and Pimental 1979: 52; Wilson and Lawrence 1985: 25; Cox 1986: 193). Indeed, so wasteful is meat production that even the present population of highly populated England could be fed from 10 million acres of the existing 46 million acres of agricultural land, if plant foods were grown and eaten directly (Williams 1977). Worldwide, over one-third of all grain is grown to feed livestock, whilst at least 500 million people are malnourished (Icke 1989: 46).

> Most of England, lowland Scotland and Wales is a shore-to-shore patchwork of fields. Most of these fields serve animal production, either as pasture or by growing fodder crops: only one field in ten produces food for people directly . . . In addition we import millions of tons of feedstuffs, much of it from developing countries. Meat production dominates agriculture worldwide: everywhere domestic animals compete with forests and wildlife and the basic needs of people. Producing and consuming less meat, fewer dairy products and eggs would release land in the developing countries and enable people there to feed themselves better.
>
> (Frances 1989: 11)

The wealthy west is seen not only to fail to share available foods equitably, but also to appropriate other people's resources – often from the poorest people of the poorest nations who can least afford it. European animal production is annually fed 22 million tonnes of imported feedstuffs, with about 15 million hectares in the 'Third World' serving as pasture for European livestock; in Senegal about 30 per cent of cultivated land performs this function; Ethiopia exports protein that would feed around one million people annually, even at the height of famine (Busacker 1985: 17; Icke 1989: 46). Vast areas of tropical forest are also felled to grow fodder crops for European and American livestock, which arouses widespread distrust:

> 'I get annoyed because McDonalds have obviously got a very busy PR department who are continually writing to papers saying "we don't use beef from tropical rainforest. Aren't we good!" – but what they don't say is where their meat does come from, or where the stuff they are fed on comes from.'

The relationships by which meat is implicated in environmentally damaging processes, with the probable consequence of further worsening food shortages for many people, can be complex. Demand for grazing land, or to grow fodder crops, to satisfy American and European craving for beef, is a principal cause of global deforestation. This demand is normally met by large commercial enterprises, often through the agency of local populations displaced from their traditional lands. Expansion of grazing is largely at the expense of forest:

> The Body Shop is asking its customers to sign letters to the president of Brazil . . . calling for action to halt the burning of tropical rainforest by ranchers seeking new pastures to produce meat for the world's hamburger chains.
>
> (Blythman 1989: 29)

> Flying over Amazonia, it seems almost inconceivable that the forests stretching for mile upon mile below, over an area almost the size of Australia, could be in jeopardy. But one has only to travel up the TransAmazonia Highway in either direction out of Altimira and the threat is all too clear. On both sides of the road, the forest has been cleared as far as the eye can see. For the most part, it has been cleared for cattle ranching. Today, there are over 8 million cattle in Brazilian Amazonia. Meat production is extremely inefficient (50kg/hectare/year), making ranching an activity which is so wholly uneconomic that it would probably never have been undertaken on the present scale if the Brazilian Government, with aid from the World Bank and other multilateral development banks, had not poured $2 billion

into subsidizing the cattle industry in Amazonia.

(Hildyard 1989: 53)

It has been calculated that when rainforest is cleared for raising cattle, the cost of each hamburger produced in the first year is about half a tonne of mature forest, since such forest naturally supports about 800,000 kilos of plants and animals per hectare, the area of which under pasture will yield some 1,600 hamburgers:

> The price of that meal-in-a-bun is anything up to nine square metres of irreplaceable natural wealth – the richness and diversity of the rainforest which may never be recreated when the grazing lands are in due course abandoned.
>
> (*New Internationalist* 1987)

Western-style intensive development also brings new problems to the areas on which it is imposed. In place of traditional indigenous forms of subsistence farming, evolved over time to support entire communities, extractive 'development' normally polarises incomes in the areas affected – in Botswana for example 75 per cent of cash income accrues to 10 per cent of the farmers (Opschoor 1985: 18) – with the less powerful losing access to food production and facing migration or starvation.

> With so much land taken up to supply our breakfast plates and steak houses, and for other cash crops, the malnourished often have to damage their own environments in an attempt to survive. Others are forced on to marginal land that is too fragile to support agriculture. It often turns to desert.
>
> (Icke 1989: 46)

Loss of primary forest also has many associated consequences such as rapid soil loss, sometimes to the point of desertification, and is implicated in global climatic change. Deforestation can reduce rainfall by 10–15 per cent in tropical areas, with frightening potential (Bunyard 1985: 1920). Some see the ecological and social devastation of economically disadvantaged nations as an early warning for catastrophic consequences of current industrial policies in the western world:

> Already local people complain of changes in the weather – the rains coming less frequently and more unpredictably.

> Indeed, scientists warn that deforestation is so disrupting the hydrological cycles which ensure the recycling of rainfall throughout Amazonia, that areas of unaffected forest downwind of deforested areas could be lost to desiccation rather than outright burning . . . The fear is that the process could go beyond the greenhouse effect . . . and actually change the chemistry of the atmosphere to such an extent that the higher mammals might not be able to survive . . . Fanciful as the idea of a climatic flip might seem, it is well to remember that for the greater part of the history of the planet, the atmosphere of the earth was such that it could only maintain bacterial forms of life.
>
> (Hildyard 1989: 59)

> [The] last thirty years have been the most disastrous in the history of most, if not all, Third World countries. There has been massive deforestation, soil erosion and desertification. The incidence of floods and droughts has increased dramatically as has their destructiveness, population growth has surged, as has urbanisation, in particular the development of vast shanty-towns, in which human life has attained a degree of squalor probably unprecedented outside Hitler's concentration camps. With such developments, have come increased malnutrition and hunger; so much so, that today we are witnessing for the first time in human history, famine on a *continental scale*, with two-thirds of African countries to some degree affected.
>
> (Goldsmith 1985: 210)

Orthodox economic experts object that such analyses are simplistic, and that meat production can serve a positive role in so-called developing countries (which are being encouraged to 'develop' on the western model, to play their allotted part in the global trade system, as sources of cheap raw materials for western markets). In Botswana for example, beef accounts for 10–20 per cent of GDP and is said to be an important source of foreign currency to a country whose environment is suggested to be suited to little else (Opschoor 1985: 18). Some draw parallels with how the English appetite for meat is said to have bailed out the poverty-stricken North of Scotland in the seventeenth-century by providing cash for a by-product of low-value pasture (Mitchison 1985: 3). To others, however, the significant parallels provided by the infamous Scottish Highland Clearances lie in the manner in which powerful élites ruthlessly evicted the indigenous population from its lands in favour of more profitable livestock for distant wealthy markets.

Meat production is also accused of being exorbitantly costly in resources and pollution. Today Europe uses only 6.5 per cent of the world's cropland to produce 28 per cent of the world's meat – seemingly excellent productivity which is bought at the expense of high resource inputs: particularly energy, metal, and phosphates. It requires around 60 gallons of water to produce a pound of wheat, but 2500 gallons of water to produce a pound of meat (Icke 1989: 46). Profit-orientation also tends to increase applications of pesticides and fertilisers and the use of modern machinery (Busacker 1985: 17–18; P. Stewart 1985: 4), with additional concerns such as the escape of chemicals into the environment, and viral immunity to antibiotics in humans and other animals due to the routine medication of livestock to promote rapid weight gain (Teherani-Kronner 1985: 12; Cox 1986: 106–108). Western resource use runs at around eight times that of poorer nations, although China is reported to achieve comparable productivity by intensive use of labour (Stewart 1985: 5).

One ecological effect of intensive farming methods that has become a political issue is nitrification of drinking water. Liquid manure from intensive animal husbandry is a major contributor. In many parts of the UK, nitrate contamination of the public water supply regularly exceeds European safety thresholds, and is reputedly linked with 'blue baby syndrome' in human infants. Nitrification is also blamed for acid rain, through a complex system of influences: nitrogen from farmland is leached into the groundwater, and presently find its way into the river networks, and from there into the seas. Observations of algal blooms in the North Sea in recent years are attributed to this increase in water fertility, the decomposition of which produces high concentrations of sulphur, which interacts with seawater and sunlight to produce atmospheric sulphuric acid. As much as 30 per cent of acid rain falling in Scandinavia may be from this source. Also linked with acid rain, and global warming, are chemicals produced by farm animals:

> Methane . . . is twenty times more efficient than carbon

> dioxide at trapping solar radiation and warming the planet . . . Mankind owns 1.2 billion head of cattle, not to mention a large number of camels, horses, pigs, sheep, and goats – together they belch about 73 million tonnes of methane into the air each year, a 435 per cent increase in the last century. The buffalo and wildebeest they displaced belched too, but their numbers were not as great.
>
> (McKibben 1990: 14)

> The main primary sources of acid rain are fumes from power stations – which can travel as much as 1,000 miles through the atmosphere – and from motor vehicles. Another significant source identified comparatively recently is ammonia produced by our burgeoning domestic livestock.
>
> (Macdonald 1990: 9)

Methane may be building up in the atmosphere at a rate of around 1 per cent per annum, with consequences impossible to compute (Bunyard 1985: 19–20).

> The 'greenhouse effect' could be avoided if we all adopted a mainly vegetarian diet, according to T.R. Vidyasagar of the Max Planck Institute for Biophysical Chemistry in Gottingen, West Germany . . .
>
> Vidyasagar calculates that a world-wide halt to the consumption of the products of grain-fed livestock, combined with the adoption of a healthy vegetarian diet, would have important consequences.
>
> With an average per capita consumption of 200 kg of grain per year, only about 60 per cent of the land now under cultivation would be needed to feed the present world population. With advancing technology, this 60 per cent should be sufficient to provide for the needs of the projected world population into the next century.
>
> Not only would this prevent further deforestation: it would allow something like 40 per cent of the present agricultural land to be reforested. As the trees grew there would be a large-scale absorption of atmospheric carbon dioxide.
>
> (*Guardian* 1987a: 24)

Expansion of beef production can produce milk disposal problems, particularly in Eastern countries where consumption of dairy products is not part of the cultural tradition (Stewart 1985: 5). Livestock production also has its effect on the appearance of the countryside. Any change in meat production levels or methods would inevitably have such consequences, just as considerable changes have already occurred throughout history. With fewer ruminants, large tracts of countryside might transfer to vegetable-growing land, forest, scrub, wildlife reserve, prairie, or recreational space (Korbey 1985: 14).

It is not only meat eating; blood sports are also accused of having various negative environmental consequences, not least of which is the sheer extent of land sometimes devoted exclusively to the maintenance of high stocks of quarry, as for example in the Scottish grouse moors. Such land, it is complained, could otherwise be put to purposes such as forestry that would be of a wider benefit to society, or be allowed to revert to a more natural state, to the benefit of wildlife. The lead shot used in shooting is also reported to pollute land, particularly in areas of high acidity such as the open moors on which shooting commonly takes place (BBC Radio 4 *PM* letters, 15 Aug. 1989). The lead weights traditionally used in fishing are similarly blamed for polluting water courses, and broken nylon fishing line threatens bird and aquatic life (*Comment*, Channel 4 TV, *c.* August 1989).

> 'Yes, I suppose it annoys me to think of so much of the country being sterilised really, so that a few wealthy people can spend their holidays blasting birds to bits.'

> Restaurants and hotels are adopting a low-flying attitude to the Glorious Twelfth this year because of fears of demonstrations by opponents of grouse shooting . . .
>
> Anti-blood-sports bodies have mounted a campaign against grouse shooting, targeting both the suffering and death of the birds and environmental pollution by lead shot.
>
> (Robertson 1989b: 3)

The meat industry, like the hunting lobby, is generally defensive on ecological issues, aware of the unpopularity of industrial farming methods. Meat producers face an urgent need to find new ways to defend their industry's environmental role. This is reflected, for example, in the trade press advertising guidebooks about the implications of the environmental movement, green issues, and promoting a green company image. To reassure the consumer about the wholesome simplicity of the product, the marketing of meat contrives to project an image of natural, timeless tradition, whilst livestock production units (previously known as farms) are shielded from the public gaze, with requests for access or information increasingly refused. A favourite strategem by anti-meat campaigners is, therefore, to make the public aware that pastoral publicity bears little resemblance to the factory conditions in which most livestock are actually reared.

Opinion is divided about the best strategy for remedying ecological problems as they become recognised. Two broad orientations emerge from the debate, with many possible positions ranged along a continuum in between.

The first, characteristically adhered to by those in positions of advantage in the status quo, for whom science, technology, and industry have brought many benefits, is that humanity's manipulation of nature, on the road to civilisation and affluence, may have produced a few unforeseen and undesirable side-effects, but that there is little or no problem that further applications of science and technology cannot resolve with sufficient attention from experts. The problem, in this conception, is insufficient technological development rather than too much.

Biotechnology, for example, is projected as the latest solution to a host of environmental problems caused by cruder technologies, promising to achieve control over rudely chaotic natural processes. Genetic engineers will, if we choose, not only design poultry with as many legs as the fried-chicken outlets require, and pigs the size of bullocks for cheaper bacon, but will even programme the beasts' genes to boost their own growth-promoting hormones to render unnecessary the administration of such drugs by costly vets. The morality of such progress is an irrelevance; the only important question is whether or not it is economically advantageous, bearing in mind consumer response.

To those of this mind, environmental problems are generally dismissed as either of minor importance or else as another exciting technological challenge. The fact that consumer expectations on one side of the world have serious social and ecological consequences for those on another part of the planet is discounted as either a matter for individual conscience, or else as an unfortunate side effect of the incontestable laws of market economics. Far from regarding it as environmentally unsound, those involved in meat production tend to see their activity as a benign use of land resources.

> 'We can convert grass into good forms of protein – i.e. red meat – and that makes sense: good economic sense. Then again, we have these vast plains and fields, and we've got to utilise grass, which is something which can be converted into meat . . . Well, I like to see things kept as naturally as possible, but you've got to look at the commercial overlay, or necessity . . . that countries must develop . . . I think that as long as there are watchdog measures that make good, common sense, then we have enough natural wit to keep the balance between commercial profit and containment actually of desirable elements in the environment . . .'

Those on the other pole of the debate typically regard the environmental problems which have increasingly come to the public's attention in recent years as inherent in industrial culture, and by definition incapable of rectification by further technological fixes. Adherents to this position often take a longer-term and more global view of the problems afflicting natural systems than is common amongst their detractors, observing, for example, that the modern environmental crisis is new only in its extent:

> The European expansionism of the last five hundred years has overshadowed – indeed, nearly totally eclipsed – the lessons that we should have learned from the repeated decline and collapse of ancient agricultural civilizations. This has led to a potentially fatal cultural blind-spot as to the vulnerability of our current industrial system of agriculture. Because of its experiences between roughly 1450 and 1950 – a period marked by seemingly unlimited expansion – the Western industrial world now finds itself conceptually ill-equipped to understand, and politically impotent to address, the problems of ecological adjustment that currently face all societies in a finite world.
>
> (Weiskel 1989: 99)

A curious example of these contradictory positions is provided by the following conversation between a livestock farmer and his wife, who between them develop many crucial features of the debate:

GERRY 'They've proved on the Nile deltas and places like that, even there they totally exhausted the natural vegetation and the soils and things, and with having slashed down the natural vegetation their ports all silted up. And you could see how things went. It's appalling. But today they have things like vast great digging machines and what have you that can go in there and un-silt rivers and things like that. That's where man has progressed. Nature really now cannot stand in his way.'

LINDA 'But maybe nature has its ways of getting its own back . . .'

GERRY '. . . well, you can talk about an earthquake. What's that, nature getting its own back?'

LINDA 'No, I'm talking about slash and burn, it's that various human beings are making money out of slashing and burning rainforest, and taking out of nature. And nature gets its own back by flooding vast areas of countryside, because the water isn't held up in the rainforest.'

GERRY 'Well, certainly if the countries haven't got the financial wealth behind them to kind of dam . . . and put banks up around the rivers – to be aware of the flooding problems. It's being sorted out here, where we're sitting now. It's prone to floods, and the last major flood here was in the fifties. But now they've spent a lot of money on it, and they've really only just completed the sea defences around the corner, but this place is now safe.'

Q. 'Even if the sea rises by several metres?'

GERRY 'Yes, I think that they've taken that into consideration in putting these sea walls in. Because you can see that the people who had the lovely sea views from the front of their houses have lost that totally now.'

Q. 'So on the whole you're fairly confident that human technology can cope with most problems as and when they arise?'

GERRY 'If it has the financial backing. Yes, certainly. Not necessarily as they arise, but given time they are aware that, like with flood plains, there's no problem in the western world with floods.'

LINDA 'Not in the western world, maybe no, but I'm talking about South America, and Africa and places like that. I mean, if there's a huge drought in Ethiopia then we can go on shipping food in for as long as we like but it's not going to cure the problem, is it? . . . And also, the other side of that, especially in Ethiopia, now that they've not got so much wood there, they dry the animal dung and burn that, which should be put on the fields to fertilise the crops, which again, you know, it's a vicious circle. There's no short term answer there. Everything has got to be very long term. I mean obviously, if there is a famine you've got to ship food in there, but that's not going to cure the problems.'

GERRY 'Then, you can say that we're the civilised world, but probably we're also the fortunate ones in being in a temperate climate, we've got . . .'

LINDA 'Well, you can thank your lucky stars that you were born . . .'

GERRY 'Yes, I know, but because things do favour us a hell of a lot more, we've managed to go streets ahead of those areas that are living in more arid areas, or . . .'

Q. 'So you'd reckon, anyway, that if Ethiopia were to become economically "developed" then they'd be able to learn to cope with these environmental problems?'

GERRY 'Well, hopefully they'd be able to get their agriculture in some sort of order then. The problem is that too many people are just living a basic living from the land. They scratch a living so there are populations that are 99 per cent farmers, or gatherers there, that live from the land. If they were economically . . . [developed, like in] Egypt perhaps – down there they are starting to do incredible things with their deserts in areas. The damming systems there they are getting now from the Nile, they are starting to bring life back to areas that were well farmed in one time. They are now starting to go back and control these areas. It's incredible that they are now starting to bring livestock back into these areas as well, although it's funny to see cattle farmed under sheltered areas, they are starting to get them back out there to get it all working again . . .'

LINDA 'Well, I think that we need to study nature and be more part of it, because with trying to control it . . . I think that most of our problems are that we try to control it, and we're just banging our heads up against a brick wall, and the more we do, the worse the problems get. We might have a small success on one hand, but then we create some dirty great other problem on the other.'

The aim here is not to say whether these relationships identified between meat production and our environment are absolutely true or false, good or bad. The significance is rather that in a remarkable number of ways, in expert discourse as in everyday communication, meat is linked to a range of ecological concerns, normally as the villain of the piece. To understand the association, we must look beyond the scientific particular, to the cultural general. As Mary Douglas notes, the growth of ecological concerns is partly a symbolic process whereby unease with anti-social behaviour – in this case possibly industrial and economic behaviour which threatens our common security – is expressed through whatever medium is immediately appropriate. The invocation of environmental pollution, as with personal health, is thus indicative of wider social concerns. Ecological concerns are valid, indeed urgent, just as meat is a nutritious substance. But the carefully proven causal relationships isolated by environmental scientists are not necessarily the most effective ways to elicit public attention:

> Another misunderstanding concerns the distinction between true and false ideas about the environment. I repeat the invitation to approach this subject in a spirit of science fiction. The scientists find out true, objective things about physical nature. The human society invests these findings with social meaning . . . In a sense the obvious risk to the environment is a distraction . . . we can never ask for a future society in which we can only believe in real, scientifically proved pollution dangers. We *must* talk threateningly about time, money, God and nature if we hope to get anything done. We must believe in the limitations and boundaries of nature which our community projects.
>
> (Douglas 1975: 242, 245–246)

But why is meat such a *focus* of environmental concerns, which by their very nature are infinitely complex systems of elements and interrelationships? Why, for example, was the contamination of lamb by radioactive fallout so particularly highlighted in the aftermath of the Chernobyl nuclear power station accident? The answer is clearly in part that meat production *does* place greater strain on the planet's resources than production of the equivalent food value from non-animal sources – on average it uses far more land, more energy, more hours of labour, and directly or indirectly it produces far more polluting wastes. And the meat from slaughtered sheep *did* contain a higher concentration of radioactive contaminant than vegetable matter by dint of being higher on the food chain.

But such rational responses do not tell the full story. Meat is also particularly indicted in such concerns since the consumption of the flesh of animals is in any case a recognised expression of our control of nature – a goal that resonates less positively in a world increasingly aware of the impact of human activity. Where meat was once an almost universally esteemed proof of human dominion over a savage and uncivilised environment, it has increasingly been represented in terms of abuse of our position of responsibility for a finite and fragile planet. Now meat represents also the fallibility of even our most advanced technologies, as exemplified by nuclear catastrophes.

> First, let us compare the ecology movement with others of historical times. An example that springs to mind is the movement for the abolition of slavery a century ago. The abolitionists succeeded in revolutionising the image of man. In the same way, the ecology movement will succeed in changing the idea of nature . . . It will succeed . . . partly because of its dedication and mostly because the time is ripe.
>
> (Douglas 1975: 231)

It is as if there are two essential strands to our thinking, each using science to understand the natural world: the first in order to overpower nature for our greater glory, and the other in order to work with nature for improvement. Today these two outlooks as to the best way forward for human society exist side by side. They are, however, mutually incompatible, and the tension between them finds expression in countless forms. One such form is our feelings about food.

As concern at the destructive excesses of industrial domination of the planet has grown, so refusal to eat the flesh of other animals has also become an ideal exemplar through which to state a preference for forms of human activity with different goals than the prosecution of an unwinnable fight to the finish against our own life support system.

15

CONCLUSIONS

This examination of meat as a social phenomenon has used two kinds of data. Firstly there is 'hard' archaeological or statistical evidence of people's eating habits, or about the medical risks and benefits of meat consumption. And secondly, there is what people say or write – their opinions. This may or may not coincide with the former sort of fact. It may seem more ephemeral, but is equally real and indispensable to understanding the first sort of evidence. Indeed, what people think to be true may if anything be *more* significant than 'facts', since belief is what governs current and future consumption. Let us review each in turn.

The first, statistical, sort of evidence is less abundant than might be supposed. We can be reasonably certain that most prehistoric peoples consumed some meat, whether hunted or farmed, just as most societies do today. The actual quantity typically eaten by our ancestors is however more difficult to ascertain. It might well have been similar to that in many modern subsistence societies where meat forms a minor proportion of the diet even though it is highly valued.

Written records – for example of taxation and of recipes – give a slightly clearer picture of eating habits during the Middle Ages, when animal flesh seems already to have been highly esteemed. At this time, for most people, meat was consumed in small quantities or only occasionally, almost as a by-product of the process of rearing animals for dairy produce and for their labour. Many modern dishes such as pizza and pasta, Irish stew and chop suey, shepherd's pie and stovies, owe their origins to traditional peasant dishes in which a large amount of a staple crop was made appetising with a little meat or vegetable flavouring.

More accurate documentary evidence of meat production and

sales is available for the early modern period. The most interesting feature is a substantial increase in meat consumption by most British people in the eighteenth and nineteenth centuries. Advances in refrigeration and transport technology in the late nineteenth century encouraged rising demand by enabling meat to be brought in cheaply from overseas. The upward trend reached a peak in the early twentieth century, after which no major further increase occurred. Wartime fluctuations obscure clear trends, but there are signs that overall demand, far from increasing, has actually begun to decline. Average figures, for example, disguise such facts as falling demand for red meats in the late twentieth century, and a steady increase in numbers of people refusing to eat meat altogether. This has been partially offset by a small further increase in consumption by some meat eaters, particularly in meals eaten away from the home, mainly in fast food establishments.

Nutrition provides few absolute facts. Much is written on the healthiness of meat eating, either extolling its virtues or deploring its debilitating consequences, but little can be said with certainty. We do know that humans have existed for many thousands of years with some animal flesh in their diets, and that some societies such as the Arctic Inuit have lived on a diet of little else. This suggests that meat cannot be quite as pernicious in terms of health as some of its more extreme detractors argue. On the other hand, there is a large body of evidence that suggests high consumption of meat – particularly of the modern intensively-reared variety – may have a price in terms of fitness and increased incidence of various degenerative disorders.

Just as significant as such empirical evidence, however, are social facts: the things which people write, say, and believe. For example, the fact that western society has traditionally used the beginnings of hunting as an indicator of the origins of humanity itself – and still (probably erroneously) characterises early human beings as primarily hunters and therefore prodigious eaters of meat – is real evidence of our modern beliefs if not of our prehistoric habits. (For example, take the vast dinosaur steaks that the amusingly anachronistic Flintstone cartoon characters guzzle.) This is but one piece in the jigsaw of evidence which suggests that meat's pre-eminence in our food system derives primarily from its tangibly representing to us the principle of human power over nature. In this case the hidden message is that we only became civilised when we began to exercise our ability to dominate other creatures by killing and eating them.

This meaning is a persistent theme which runs through context after context associated with meat in western society. The depth to which our aversion to cannibalism extends, to the point that many of us would literally prefer to die than 'descend' to eating human flesh, illustrates the absolute distinction which we have traditionally drawn between inedible people and edible non-human animals. Our aversion to eating pets, carnivores, or other creatures which we see as being too close to ourselves conforms to the same distinction.

The notion of environmental control also provides the context for the rational reasons to which we impute our carnivorism, or our distaste for meat. The value judgements which underpin our economic system regulating meat production and consumption are strongly governed by this idea. So are our beliefs, reinforced by medical specialists, as to whether or not it is healthy for us in body and in mind to eat much meat. Our ethical principles concerning the proper treatment of non-human animals, including the justice of consuming them for food, are also clearly influenced by our view of the correct relationship of humans to the environment in general; and this is likewise a central theme in the modern debate on ecological threats to our continued existence as a species, in which meat is regularly implicated.

A picture emerges of meat as a symbol by which western society – like many other societies – has long expressed its relationship to the world that it inhabits. Through most of our history, and presumably prehistory, people have experienced a need to control their environment: to mitigate the threat from the elements and from wild animals, and to ensure some stability in the supply of food and other necessities of life. In this context, it seems appropriate that for most subsistence societies, and certainly in medieval Europe, the prevailing use of meat appears to have been as an esteemed supplement to the basic diet of grains and vegetables. Principal exceptions to this rule have been those who particularly sought political and economic power, who seem to have consumed meat in greater than average quantities, and those who shunned earthly power for spiritual reasons and accordingly shunned meat as well.

It is also fitting that, in most people's diets, meat should have risen in both quantity consumed and in significance from around the seventeenth century onwards, at a time when science was increasingly stressing the need to dominate nature, morally abetted by mechanistic philosophers who portrayed non-human animals as little more than sophisticated machines. Whereas environmental control might previously have been a basic necessity for most people making a living from the land, to an unprecedented extent it now became an ethical imperative for urban society. Meat provided the ideal expression by which the power of human industry could be demonstrated.

However, even as this new ethic was rising to prominence, some were also rejecting the tenets of that prevailing philosophy. As urban society came to have less and less daily contact with the environment on which it ultimately depended, a significant minority of people became concerned at excessive abuses of human power. The eighteenth and nineteenth centuries saw the foundation of nature reserves, of societies for the protection of animals, and of vegetarian movements – as refusal to eat meat on moral grounds became no longer the preserve of the devout.

In the twentieth century the tension between these alternative positions has continued, and their relative influences go a long way towards explaining our modern views of meat. If anything, indeed, the arguments have fragmented and polarised, with the growth of extreme militancy in defence of animals on the one hand, and on the other the intensive industrialisation of meat supply whereby animals are reared from conception to slaughter on a production line which makes few concessions to their creature comforts.

As yet, no single alternative philosophy has emerged to challenge the exploitative industrial paradigm – instead a loosely knit cluster of viewpoints has been coalescing under a green umbrella. These include, for example, 'light green' advocates of such concepts as 'stewardship', who maintain a securely anthropocentric stance, but who suggest that humans must take much better care of the planet that we govern. There are also the 'deep greens': those who believe that the distinction that places humans apart from, and above, the rest of nature is a fundamental fallacy and must be shed, so that we learn to value the planet's interests as, literally, equivalent to our own.

Amongst those avoiding the consumption of meat there is a related diversity of views. Many believe eating the flesh of other animals to be unnecessary, unhealthy, unjust, barbaric, and therefore utterly wrong. Not every vegetarian holds such an extreme opinion, however. Many do not find the consumption of meat, as such, abhorrent, but refuse to condone the industrialised treatment of animals and will not, therefore, consume its product. Others find abrogation of responsibility for killing objectionable, but say they would happily eat the flesh of animals that they had killed for themselves. Then there are the many who have moved towards low-meat diets – with a preference for organic and free-range meat. Others state that they would eat a certain amount of meat if the animals were more humanely treated, but will continue to avoid it altogether so long as most meat production is carried out on the current industrial model.

The answer to my initial question – 'why do we value animal flesh so highly, in spite of the consequences for the creatures involved?' – has, in effect, been that we do not esteem meat *in spite* of the domination of sentient beings. Rather, excepting the qualms that we may (individually) feel when faced with our responsibility for a living animal's death, we (as a society) esteem meat so highly partly *because* of that power. It is not that we each consciously exult in our mastery of nature whenever we bite into a piece of flesh, but we are brought up within a culture which has regarded environmental conquest as a laudable goal, and which has deployed meat as a primary means to demonstrate it.

Whilst this discussion has centred upon the ways in which we think about meat, the analysis holds further implications for anyone who wishes to consider them. The wearing of animal furs, for example, can be seen as an analogous exhibition of human authority. More controversially perhaps, much of the edifice of physiological experimentation and product testing on animals could be said to exhibit similar traits. Many such procedures are, after all, neither necessary nor reliable. It may not be going too far to suggest that the development of vivisection has never merely been about obtaining scientific data – it may also have functioned to gratify the positivistically-inclined establishment's desire to display its mastery of unruly nature.

No great feat of imagination is required to perceive the ideology of power as a persistent thread woven throughout the fabric of our lives. Whether expressed as 'achievement', 'freedom', 'convenience', or 'prestige', a desire to transcend the limits of nature pervades western beliefs and values. The modern devotion to the motor car, for example, owes much to the machine's ability to enhance individual mobility. The example below, however, illustrates how readily such an unimpeachable aspiration can become perverted. The car has become not just a means to mobility, but a medium through which to express the capacity to move further, faster, in more comfort, or merely at greater expense *than others*. In the words of an American executive:

> When I get into the limo on a rainy evening in December, and look out the window, it isn't speed that I think about. For God's sake, the subway would be faster. No, I say to myself, 'Those people out there are getting cold and wet, and I'm in here warm and dry.' I had to go through a lot of shit to get where I am, but when I look at people waiting for a bus in the rain, it makes it all worthwhile. They know I've made it. I know I've made it. You can't beat a limo for that.
>
> (Korda 1975: 213)

Leaving aside the spiritual desirability of the ideology illustrated here, defining success as fulfilment gained through power over others also carries with it the inevitability of losers. The quest for power can have serious consequences. Slavery, for example, accompanied 'economic development' by the wealthy nations of the west. And, once again, its value to slave owners probably lay not only in the physical labour of the human beings enslaved: the very act of owning other humans also testifies to the power of the owner. A modern corollary, the economic exploitation of the 'Third World' by industrially developed nations, may conceivably serve a similar function. It is not a pleasant idea, but perhaps the extreme poverty that western wealth causes in the Third World is even of some obscurely corrupt importance to us: a subliminal reassurance that we really are well off.

This is not, of course, to imply that such implications are necessarily a source of conscious satisfaction. I have sought to stress that such ideas need seldom, if ever, be directly thought in order to be significantly influential. They can operate at a level of cultural consensus rather than individual awareness, and may be all the more powerful for that. The individual male need not deliberately subscribe to an explicit manifesto of male supremacy over women to be swayed by, and to contribute to, a social system in which female's opportunities are severly limited. So our communal value of power over nature has substantially supported the emphatic prestige of meat, even though this notion may have been specifically elaborated only rarely, by a few scientists and philosophers.

Western society's pursuit of a prosperity defined by advantage over others also, however, has a conspicuous tendency to be self-defeating. As more and more people seek the prestige of individually motorised mobility, each individual becomes less mobile, ensnarled in traffic chaos. Racism, as exemplified by slavery and the haemorrhaging of 'Third World' nations' wealth, has bequeathed a perilous legacy of social and political hostilities. And the environmental catastrophe precipitated by our culture's attempt to exert its sacred authority over brute nature has, it is now clear, left human civilisation itself teetering on a precipice to oblivion.

The meat industry is currently in a state of considerable uncertainty. Many producers regard themselves as almost under siege, on account of the rapid changes which have occurred in British consumer preferences in recent years. At time of writing, British beef sales are reported to have, at least temporarily, plummeted by up to 50 per cent as the public has learned of the potential implications of epidemic BSE. Some experts talk ominously of unimaginable consequences for human health or for the trade, whilst government spokespeople characteristically seek to defend the industry with 'scientific' reassurances. These conflicting messages in part reflect alternative views of the meat system, which correspond to alternative views of the nature of human society.

On the one hand there is the radical view that meat eating is inherently either unhealthy, or unethical, or both. On the other, the orthodox view is still that meat is nutritious and even necessary, except when it is too high in saturated fat content or is excessively contaminated by BSE, salmonella, listeria, BST, hormone residues, antibiotic residues, or whatever the latest concern may be. The picture, however, is not so simple.

Over-the-counter sales of red meat – the epitome of meat – have fallen steadily over the past couple of decades in particular. Increasingly, producers have had to divorce their products from associations with the flesh of real, live animals in order to maintain customer acceptability, particularly among the younger generation. A plethora of prepared and processed products is the result. In the process, however, the industry may well have sown for itself the seeds of an even greater problem. Heavily advertised 'coated nuggets' and exotic vacuum-packed dishes have persuaded consumers to continue buying meat in various new forms, but have also consolidated many people's disinclination to deal directly with raw flesh. The danger for meat producers is that there may be little further potential for disguising the product, and little prospect of convincing an increasingly squeamish public to return to the old ways. Meat sales are dependent upon those whose faith not just in the meat industry, but in the benefits of the industrial economy in general, remains largely untarnished. Should that faith be undermined the food industry in general, and meat industry in particular, may be in a vulnerable position.

The sting in the tail is that a collapse of consumer confidence in the products of the industrial food industry looks increasingly possible. Meat sales, particularly of the most emotive red meats, have clearly already suffered by being associated as a symbol of human domination of the planet in general – a process which is seen by many as having gone too far. The rapid growth in environmental concern in recent decades, and a parallel growth in the number of people choosing to eat less meat or no meat at all, has certainly been stimulated by the unmistakable evidence of severe environmental (and consequently social) damage in countless areas together with a growing awareness of the possibility of imminent ecological catastrophe on an unprecedented scale. But in a world locked into a seemingly unbreakable cycle of economic growth, defined by ever-increasing production and consumption of material resources, there is every likelihood that environmental crises may become more obvious, more frequent, and more severe in the foreseeable future.

There are two typical responses to environmental crises. One is to regard individual ecological problems in isolation as the result of inadequate scientific understanding and poor control, and to seek to rectify them by further applications of industrial technology. This is the characteristic response of those who adhere to the tenets of recent western industrial culture, who view the world as an infinite resource and a challenge to be overcome. In this view, archetypically, humans are a species set apart, unconstrained by the physical limitations of other animals, and unique in our capacity to modify the world to the blueprint of our choice. Such true believers typically have faith in the ability of science and technology to find solutions to any and every problem, whilst continuing to provide the material comforts to which we in the west have become accustomed. For such people, who remain the majority, meat continues to fulfil its traditional function of exemplifying that value of human pre-eminence and – health scares apart – remains popular.

At the other pole of opinion are those who regard the environmental crises as inherent in our current cultural constitution, who see individual ecological problems not in isolation but as interrelated symptoms of a wider malaise. Whatever their particular vision of the future, most such individuals believe that only by adopting a more sensitive approach to our dealings with the planet – including recognition that the non-human environment has needs which must sometimes override our immediate demands – can catastrophic deterioration in local and global ecosystems be averted. It is in these circles that the reputation of meat, as a continuing symbol of human domination of nature, has suffered most severely.

There is a tension between two broad orientations: the industrial and the ecological (each of which, upon closer examination, is fragmented into a kaleidoscopic diversity of viewpoints). This tension can only heighten in the future, as threats to the global environment and consequently to humans' quality of life, worsen. The effects of social attitudes more sympathetic towards the needs of the non-human world are already visible – not merely in the rising popularity of vegetarianism but also, for example, in increased demand for organic and free-range meats; in the trend away from 'red' meats towards 'white'; in demands for more humane and natural treatment of farmed animals; in militancy towards practices regarded by some as unnecessarily cruel such as hunting and vivisection; in the purchase of meat in increasingly disguised forms, such as hamburgers or prepared meals; and – not least – in the new public perception of all meat as slightly unhealthy in various ways, in contrast to its traditional image as essential, vital nutrition.

Should the consensus of opinion in future society dictate that nature must be dealt with more sensitively, meat may well continue to be used as an expression of our relationship to our environment, and its social acceptability fall as a consequence. It is at least possible that, in this way, in some years time meat eating could come to have a widespread image comparable to that of, say, smoking or drug addiction today – as a relatively vulgar, unhealthy and anti-social indulgence.

History shows that public values can and do change. For example, a practice such as slavery which was once generally acceptable and regarded as entirely normal is now widely regarded with horror – as a sign not of high civilisation, but of barbaric brutality. There is no self-evident reason that our consumption of the flesh of other animals should be immune from a similar process. Since 'carnivoracity' has long been a Natural Symbol by which we have expressed our society's quest for dominance, the food's diminishing status could well be symptomatic of the wane of outdated ideals. If so, the turbulently declining reputation of meat, at the advent of the third millenium, may be a harbinger of the evolution of new values.

BIBLIOGRAPHY

Achebe, C. (1988) 'An Image of Africa: Racism in Conrad's "Heart of Darkness" ', in *Hopes and Impediments: Selected Essays* 1965–1987: 1–13, London: Heinemann.

Adair, J. (1775) *A History of the American Indians,* London: Dilly.

Adams, C.J. (1990) *The Sexual Politics of Meat,* New York: Continuum.

Adams, D. (1988) *The Long Dark Tea-Time of the Soul,* London: Pan.

Adams, R. (1989) 'Untender traps and fatal attractions', *Guardian,* 28 January: 9.

Addis, W. and Arnold, T. (eds) (1924) *A Catholic Dictionary,* 9th edn, London: Virtue & Co.

Andrewes, L. (1650) *A Pattern of Catechistical Doctrine,* 1846 edn, Oxford.

Angyal, A. (1941) 'Disgust and Related Aversions', *Journal of Abnormal and Social Psychology* 36.

Ardener, E. (1971) 'Introductory Essay', in *Social Anthropology and Language,* ASA Monographs 10: ix–cii, London: Tavistock.

—— (1975) 'Belief and the Problem of Women', in S. Ardener (ed.) *Perceiving Women*: 1–17, London: Dent.

Arens, W. (1979) *The Man-Eating Myth: Anthropology and Anthropophagy,* New York: Oxford University Press.

Aristotle (1984) 'Politics', in J. Barnes (ed.) *The Complete Works of Aristotle: The Revised Oxford Translation,* Guildford, Surrey: Princeton University Press, pp. 1993–1994

Armstrong, B.K., van Merwick, A.J., and Coates, H. (1977) 'Blood Pressure in Seventh-day Adventist Vegetarians', in *American Journal of Epidemiology* 105: 444–449.

Associated Press (1989) 'Anti-meat activists raiding ranches', *San Francisco Sunday Examiner and Chronicle,* 26 November: B7.

Atwood, M. (1969) *The Edible Woman,* 1979 edn, London: Virago.

Bach, R. (1985) *The Bridge Across Forever,* London: Pan.

Back, K. (1977) 'Food, Sex and Theory', in T.K. Fitzgerald (ed.) *Nutrition and Anthropology in Action*: 24–34, Amsterdam: Van Gorcum.

Baker, R. (1973) 'Red meat decadence', *New York Times,* 3 April: 43.

Barthes, R. (1975) 'Towards a Psychosociology of Contemporary Food Consumption', in E. Forster and F. Forster (eds) *European Diet, from*

Pre-Industrial to Modern Times: 47–59, New York: Harper & Row.

Bennet, M.K. (1954) *The World's Food,* New York: Harper & Row.

Bentham, J. (1789) *Introduction to the Principles of Morals and Legislation,* London: T. Payne.

Berlin, B. and Kay, P. (1969) *Basic Color Terms: Their Universality and Evolution,* Berkeley: University of California Press.

Bernard, T. (1982) *Hatha Yoga,* London: Rider.

Bingley, W. (1824) *Animal Biography, or Popular Zoology Illustrated by Authentic Anecdotes,* 6th edn, London: C. & J. Rivington.

Birke, L. (1986) *Women, Feminism and Biology: The Feminist Challenge,* Brighton: Wheatsheaf.

Bloch, M. (1985) 'Religion and Ritual', in A. Kuper and J. Kuper (eds) *The Social Science Encyclopedia,* London: Routledge & Kegan Paul.

Blythman, J. (1989) 'Briefing', *Scotland on Sunday,* 23 July: 30.

Bold, A. (ed.) (1980) *Mounts of Venus: The Picador Book of Erotic Prose,* London: Pan.

Bourdieu, P. (1977) *Outline of a Theory of Practice,* Cambridge: Cambridge University Press.

Bowle, J. (1979) *A History of Europe,* London: Pan.

Bowser, D. (1986) 'Stalking, Past and Present', in *The Changing Scene,* proceedings of the 1986 Deer Conference organised by the British Deer Society and the Red Deer Commission: 22–24.

Braudel, F. (1974) *Capitalism and Material Life,* 1400–1800, London: Fontana.

British Meat (1987a) 'Minister congratulates British meat industry on record exports – opportunities for further expansion ahead', Milton Keynes: Meat and Livestock Commission, Summer: 2–3.

—— (1987b) 'Commissioners urge product improvement', Milton Keynes: Meat and Livestock Commission, Summer: 4.

Brody, J. (1981) *Jane Brody's Nutrition Book,* New York: W.W. Norton.

Brown, J.E. (1972) *The North American Indians,* New York: Aperture.

Brown, P. and Jordanova, L.J. (1982) 'Oppressive Dichotomies: The Nature/Culture Debate', in The Cambridge Women's Studies Group (ed.) *Women in Society:* 224–241, London: Virago.

Brownmiller, S. (1975) *Against Our Will: Men, Women and Rape,* New York: Simon & Schuster.

Bunyard, P. (1985) 'Possible Consequences of Increased Rice Growing and Cattle Raising in the Tropics', in Centre for Human Ecology (ed.) *Focus On Meat*: 19–20, transcript of proceedings of European Workshop in Human Ecology, 27–31 May, University of Edinburgh.

Burnett, J. (1966) *Plenty and Want: A Social History of Diet in England from 1815 to the Present Day,* London: Nelson.

Busacker, D. (1985) 'Satisfied Livestock, Degraded Land and Hungry People: Feedstuff Imports as a Barrier to an Ecological Society', in Centre for Human Ecology (ed.) *Focus On Meat*: 17–18, proceedings of European Workshop in Human Ecology, 27–31 May, University of Edinburgh.

Butcher & Processor (1987a) Speed butchery equipment advertisement, London: Smithfield Publishing, November: 15.

Butcher & Processor (1987b) 'More help for trade in new MLC plan', London: Smithfield Publishing, November: 5.

—— (1987c) 'Colin likes to keep his business traditional', London: Smithfield Publishing, December: 6.

—— (1989a) 'Supermarket giants still big threat to butchers', London: Smithfield Publishing, April: 5.

—— (1989b) 'Red packs quite a punch', London: Smithfield Publishing, May: 19.

Carroll, L. (1872) 'Through the Looking Glass: and What Alice Found There', in *The Penguin Complete Lewis Carroll,* 1982 edn, Harmondsworth: Penguin.

Cecil, D. (1929) *The Stricken Deer; or, The Life of Cowper,* London: Constable.

Central Statistical Office (1990) *Annual Abstract of Statistics,* London: HMSO.

Centre for Human Ecology (ed.) (1985) *Focus on Meat,* transcript of proceedings of European Workshop in Human Ecology, 27–31 May 1985, University of Edinburgh: Centre for Human Ecology.

Chagnon, N. (1977) *Yanomamo, The Fierce People,* London: Holt, Rinehart & Winston.

Chaitow, L. and Martin, S. (1988) *A World without AIDS,* Wellingborough: Thorsons.

Chiltosky, M.U. (1975) 'Cherokee Indian Foods', in M. Arnott (ed.) *Gastronomy: The Anthropology of Food and Food Habits*: 235–244, The Hague: Mouton.

Chou, M. (1979) 'The Preoccupation with Food Safety', in M. Chou and D.P. Harmon (eds) *Critical Food Issues of the Eighties*: 18–41, New York: Pergamon.

Clayton, H. (1978) 'Soya producers fight on', *The Times,* 24 November: 6.

Cockayne, O. (1864) *Leechdoms, Wortcunning, and Starcraft of Early England,* London: Longman.

Cohen, A. (1986) 'Of Symbols and Boundaries, or, Does Ertie's Greatcoat Hold the Key?', in A. Cohen (ed.) *Symbolising Boundaries: Identity and Diversity in British Cultures*: 1–19, Manchester: Manchester University Press.

Cohen, N. (1987) 'Chilling fiction gives meat trade a novel boon', *Independent,* 22 December: 1.

Cohen, P. (1967) 'Economic Analysis and Economic Man: Some Comments on a Controversy', in R. Firth (ed.) *Themes in Economic Anthropology,* ASA Monographs series 6: 91–118, London: Tavistock.

Collier, J. and Rosaldo, M. (1981) 'Politics and Gender in Simple Societies', in Ortner and Whitehead (eds) *Sexual Meanings:* 275–329, London: Cambridge University Press.

Comfort, A. (1974) *Joy of Sex. A Gourmet Guide to Lovemaking,* London: Quartet.

Coon, C. (1955) *The History of Man: From the First Human to Primitive Culture and Beyond,* London: Cape.

Cooper, C., Wise, T. and Mann, L. (1985) 'Psychological and Cognitive Characteristics of Vegetarians', *Psychosomatics* 26, 6: 521–527.

Co-operative Women's Guild (1978) *Maternity: Letters from Working Women,* London: Virago.
Cowart, B.J. (1981) 'Development of Taste Perception in Humans: Sensitivity and Preference Throughout the Life Span', *Psychological Bulletin* 90, 1: 43–73.
Cox, M. and Crockett, D. (1979) *The Subversive Vegetarian: Tactics, Information and Recipes for the Conversion of Meat Eaters,* Wellingborough: Thorsons.
Cox, P. (1986) *Why You Don't Need Meat,* Wellingborough: Thorsons.
Crawford, R. (1985) 'A Cultural Account of "Health": Control, Release and the Social Body', in J. McKinley (ed.) *Issues in the Political Economy of Health Care*: 60–103, London: Tavistock.
Crawshaw, S. (1987) 'Meat eating in the States', *The Independent,* 5 May: 9.
Culler, J. (1976) *Saussure,* The Hassocks, Sussex: Harvester.
Daily Telegraph (1988) 'Cruelty' label for fur coats, 8 February: 1.
Dallas, E. (1877) *Kettner's Book of the Table,* London: Dulau.
Dalton, G. (1961) 'Economic Theory and Primitive Society', *American Anthropologist* 63: 1–25.
Dando, W.A. (1980) *The Geography of Famine,* London: Arnold.
Darlington, C. (1969) *The Evolution of Man and Society,* New York: Simon & Schuster.
Davies, S. (1988) 'Meal on a string', *The Guardian,* 12 July: 34.
Davis, H. (1946) *Moral and Pastoral Theology* II, London: Sheed & Ward.
Doddridge, P. (1763) *A Course of Lectures on the Principal Subjects in Pneumatology, Ethics and Divinity,* London: Robinson.
Douglas, M. (1966) *Purity and Danger,* London: Routledge & Kegan Paul.
—— (1970) *Natural Symbols,* London: Barrie & Rockliff.
—— (ed.) (1973) *Rules and Meanings,* Harmondsworth: Penguin.
—— (1975) *Implicit Meanings,* London: Routledge & Kegan Paul.
—— (1978) *Culture,* in Annual Report of the Russell Sage Foundation: 55–81, New York.
Douglas, M. and Isherwood, Baron (1980) *The World of Goods,* Harmondsworth: Penguin.
Douglas, M. and Nicod, M. (1974) 'Taking the Biscuit: The Structure of British Meals', *New Society* 19: 744–747.
Drowser, E. (1937) *The Mandaeans of Iraq and Iran: Their Cults, Customs, Magic, Legends, and Folklore,* Oxford: Clarendon Press.
Dulverton, Lord (1986) 'Introduction' to *The Changing Scene,* proceedings of the 1986 Deer Conference organised by the British Deer Society and the Red Deer Commission: 5–6.
Dumont, L. (1972) *Homo Hierarchicus,* London: Paladin.
Dwyer, J.J., Mayer, R.F., Kandel, R.F. and Mayer, J. (1973) *Journal of American Diet Association* 62: 503.
Eagle, R. (1978) 'Hidden hazards in the hamburger?', *Sunday Times,* 17 September: 12, reporting Science 210, 4359.
Eckstein, E.F. (1980) *Food, People and Nutrition,* Westport: AVI.
Edinburgh Evening News (1990) 'Eating out', 5 January: 5.
Ehrlichman, J. (1990) 'Meat eaters swallow food poison risk', *Guardian,* 29 January: 3.

Eisenstein, H. (1984) *Contemporary Feminist Thought*, London: Unwin.

Elias, N. (1939) *The Civilising Process*, 1978 edn, New York: Urizon.

Engels, F. (1844) 'The Condition of the Working-Class in England', in *Karl Marx, Frederick Engels: Collected Works* IV, 1975 edn, London: Lawrence & Wishart.

Erhard, D. (1973) 'The New Vegetarians', *Nutrition Today* 8: 4–12.

Evans, E.P. (1987) *The Criminal Prosecution and Punishment of Animals*, London: Faber & Faber.

Evans, M. (ed.) (1982) *The Woman Question*, London: Fontana.

Feirstein, B. (1982) *Real Men Don't Eat Quiche*, London: New English Library.

Fenton, A. and Kisbán, E. (eds) (1986) *Food in Change: Eating Habits from the Middle Ages to the Present Day*, Edinburgh: John Donald.

Fiddes, N. (1989) *Meat: A Natural Symbol*, unpublished PhD thesis, University of Edinburgh.

Fieldhouse, P. (1986) *Food and Nutrition: Customs and Culture*, London: Croom Helm.

Flowers, C. (1989) 'Language of the interview', *Scotland on Sunday*, 25 June: 19.

Fox, R. (1985) 'The Conditions of Sexual Evolution', in P. Ariès and A. Béjin (eds) *Western Sexuality: Practice and Precept in Past and Present Times*: 1–13, Oxford: Blackwell.

Frances, M. (1989) *Small Change: a Pocketful of Practical Actions to Help the Environment*, University of Edinburgh: Centre for Human Ecology.

Frank, J. (1987) *Dietary Trends in the United Kingdom*, University of Bradford: Food Policy Research Unit.

Gandhi, M. (1949) *Diet and Diet Reform*, Ahmedabad: Navjivan.

George, S. (1984) *Ill Fares the Land: Essays on Food, Hunger, and Power*, Washington: Institute for Policy Studies.

Goldsmith, E. (1985) 'Is Development the Solution or the Problem', *The Ecologist* 15, 5/6: 210–219.

Goodall, J. (1965) 'Chimpanzees on the Gombe Stream Reserve', in I. DeVore (ed.) *Primate Behavior*, New York: Holt, Rinehart & Winston.

Goody, J. (1982) *Cooking, Cuisine and Class*, Cambridge: Cambridge University Press.

Graham, T. (1835) *Modern Domestic Medicine*, London: Simpkin & Marshall.

Greene, K. (1986) *The Archaeology of the Roman Economy*, Berkeley: University of California Press.

Griffin, S. (1981) *Pornography and Silence: Culture's Revenge against Nature*, London: Women's Press.

Griggs, B. (1986) *The Food Factor*, Harmondsworth: Viking.

Gross, D. (1975) 'Protein Capture and Cultural Development in the Amazon Basin', *American Anthropologist* 77: 526–549.

Guardian (1987a) 'Eating our way out of the greenhouse effect', 1 May: 24.

—— (1987b) 'Moscow to get taste of Big Mac', 21 November: 6.

Guy-Gillet, G. (1981) 'La cucina nel campo analitico', *Rivista-di-Psicologia-Analitica* 12, 23: 114–128.

Hager, C. (1985) *Demand for nutrient and non-nutrient components in household purchases of red meat, poultry, and fish products using a hedonic approach (theory, consumer demand, implicit prices, food consumption)*, unpublished PhD thesis, North Carolina State University.

Hammet, R.C. and Nevell, W.H. (1929) *A Handbook on Meat and Textbook for Butchers*, London: The Meat Trades Journal Co. Ltd.

Hardinge, M.G. and Crooks, H. (1964) 'Non-Flesh Dietaries, 3: Adequate and Inadequate', *Journal of the American Dietetic Association* 45: 537.

Harlan, J.R. (1976) 'The Plants and Animals that Nourish Man', *Scientific American* 235, 3: 88–97.

Harrington, G. (1985) *Meat in the Modern World*, Milton Keynes: Meat & Livestock Commission.

Harris, M. (1975) *Cows, Pigs, Wars and Witches*, New York: Vintage.

—— (1986) *Good to Eat*, London: Allen & Unwin.

Harvey, B. and Hallett, J. (1977) *Environment and Society*, London: Macmillan.

Haudricourt, A. (1962) 'Domestication des animaux, culture des plantes et traitement d'autrui', in *L'Homme* II.

Health & Strength Magazine (1901) 'About eating' 6, 3: 358–359.

Hebert, H. (1987) 'The naked and the dead', *Guardian*, 31 December: 17.

Hegsted, D.M., Trulson, M.F., White, H.S., White, P.L., Vinas, E., Alvistur, E., Diaz, C., Vasquez, J., Loo, A., Toca, A., Collazos, C., and Ruiz, A. (1955) 'Lysine and Methionine Supplementation of All-Vegetable Diets for Adult Humans', *Journal of Nutrition* 56: 555.

Henderson, H., Lintott, J., and Sparrow, P. (1986) 'Indicators of No Real Meaning', in P. Ekins (ed.) *The Living Economy*: 32–39, London: Routledge & Kegan Paul.

Hildyard, N. (1989) 'Adios Amazonia? A Report from the Altimira Gathering', *The Ecologist* 19, 4: 53–62.

Hill, G.B. (ed.) (1964) *Boswell's Life of Johnson*, Oxford: Oxford University Press.

Hopkins, E. (1692) *A Practical Exposition on the Ten Commandments.*

Horigan, S. (1988) *Nature and Culture in Western Discourses*, London: Routledge.

Hughes, J.D. (1975) *Ecology in Ancient Civilizations*, Albuquerque: University of New Mexico Press.

Icke, D. (1989) 'Eaten out of house and home', *The Times*, 9 September 1989: 46.

IFAW (1988) 'Every Time You Hear Korean Olympics Think of This!', International Fund for Animal Welfare advertisement, *Guardian*, 27 May: 3.

Illich, I. (1976) *Limits to Medicine. Medical Nemesis: The Expropriation of Health*, Harmondsworth: Penguin.

Independent (1989) 'Quote unquote', 6 May: 16.

Ingold, T. (1986) *The Appropriation of Nature*, Manchester: Manchester University Press.

Jackman, B. (1989a) 'The price of meat', *Sunday Times Magazine*, 12 November 1989: 46–50.

—— (1989b) 'Making a killing', *Sunday Times Magazine*, 12 November 1989: 50–52.

Jacobs, A.H. (1958) 'Masai Age-groups and some Functional Tasks', East African Institute of Social Research Conference Proceedings, Makerere.

Jancey, M. (1987) *Mappa Mundi,* Leominster, Herefordshire: Orphans Press.

Jochelson, W. (1908) *The Koryak,* Jesup North Pacific Expedition VI, American Museum of Natural History Memoir 10, Leiden: E.J. Brill.

Keating, F. (1988) 'From a view to a death in the morning', *Guardian,* 27 December 1988: 17.

Kenton, L. and Kenton, S. (1984) *Raw Energy,* London: Arrow.

Kephart, J. (1970) *Primitive Woman as Nigger; or, The Origin of the Human Family as Viewed Through the Role of Women,* unpublished M.A. Dissertation, University of Maryland.

Kerr, M. and Charles, N. (1986) 'Servers and Providers: the Distribution of Food within the Family', *Sociological Review* 34, 1: 115–157.

King, W. (1731) *An Essay on the Origin of Evil.*

Kisbán, E. (1986) *Food Habits in Change: The Example of Europe,* in Fenton and Kisbán (eds.) *Food in Change: Eating from the Middle Ages to the Present Day:* 2–10, Edinburgh: John Donald.

Klingender, F. (1971) *Animals in Art and Thought, to the End of the Middle Ages,* E. Antal and J. Harthan (eds.), London: Routledge & Kegan Paul.

Kolakowski, D. and Malina, R.M. (1974) 'Spatial Ability, Throwing Accuracy and Man's Hunting Heritage', in *Nature* 251, 5474: 410–412.

Korbey, A. (1985) *Food Production and Our Rural Environment – The Way Ahead,* University of Reading: Centre for Agricultural Studies.

Korda, M. (1975) *Power: How to Get it, How to Use it,* New York: Ballantine.

Lancaster, J. (1968) 'The Evolution of Tool-using Behavior: Primate Field Studies, Fossil Apes and the Archaeological Record', *American Anthropologist* 70: 56–66.

Land, H. (1969) *Large Families in London,* London: Bell.

Leach, E. (1964) 'Anthropological Aspects of Language: Animal Categories and Verbal Abuse', in P. Maranda (ed.) *Mythology* 1972 edn: 39–67, Harmondsworth: Penguin.

—— (ed.) (1967) *The Structural Study of Myth and Totemism,* ASA Monographs 5, London: Tavistock.

—— (1970) *Claude Lévi-Strauss,* New York: Viking.

—— (1974) *Lévi-Strauss,* revised edn, London: Fontana.

Lecky, W.E.H. (1869) *History of European Morals from Augustus to Charlemagne,* 1913 edn, London: Watts.

—— (1955) *History of European Morals,* New York: Brazilier.

Lee, R.B. (1972) 'The !Kung Bushmen of Botswana', in M. G. Bicchieri (ed.) *Hunters and Gatherers Today,* New York: Holt, Rinehart & Winston.

Le Gros Clark, F. (1968) 'Food Habits as a Practical Nutrition Problem', *World Review of Nutrition and Dietetics* 9: 56–84.

LeMay, L. and Zall, P.M. (eds) (1986) *Benjamin Franklin's Autobiography,* New York: Norton.

Le Roy Ladurie, E. (1978) *Montaillou,* 1978 edn, Harmondsworth: Penguin.

Lévi-Strauss, C. (1963) *Structural Anthropology,* London: Basic Books.

Lévi-Strauss, C. (1966) 'The Culinary Triangle', *New Society* 166: 937–940.
—— (1969) *The Elementary Structures of Kinship,* London: Eyre & Spottiswoode.
—— (1970) *The Raw and the Cooked,* London: Cape.
—— (1973) *From Honey to Ashes,* London: Cape.
—— (1978) *The Origin of Table Manners,* London: Cape.
—— (1987) *Anthropology and Myth,* Oxford: Blackwell.
Liebig, Baron Justus von (1846) *Animal Chemistry,* 3rd edn, London: Taylor & Walton.
—— (1847) *Researches on the Chemistry of Food,* London: Taylor & Walton.
Linzey, A. (1985) *The Status of Animals in the Christian Tradition,* Birmingham: Woodbroke College.
Lovejoy, A. (1936) *The Great Chain of Being,* Cambridge, Massachusetts: Harvard University Press.
McCormick, F. (1987) *Stockrearing in early Christian Ireland,* unpublished PhD thesis, Queen's University of Belfast.
McCurdy, E. (1932) *The Mind of Leonardo da Vinci,* London: Cape.
Macdonald, P. (1990) 'Just how close is breaking point?', *The Scotsman,* 13 January: 9.
McElroy, A. and Townsend, P. (1985) *Medical Anthropology in Ecological Perspective,* London: Westview Press.
McGuinness, D. (1976) 'Away from a Unisex Psychology: Individual Differences in Visual Sensory and Perceptual Processes', *Perception* 5: 279–294.
McKibben, B. (1990) *The End of Nature,* London: Viking.
McLaughlin, T. (1978) *A Diet of Tripe: The Chequered History of Food Reform,* Newton Abbot, Devon: David & Charles.
MAFF (Ministry of Agriculture, Fisheries and Food) (1982) Annual Report of the National Food Survey Committee, London: HMSO.
Majumder, S.K. (1972) 'Vegetarianism: Fad, Faith, or Fact?', *American Scientist* 60, 175: 175–179.
Margetts, B.M., Beilin, L.J., Vandongen, R., and Armstrong, B.K. (1986) 'Vegetarian Diet in Mild Hypertension: A Randomised Controlled Trial', *British Medical Journal* 293, 6560: 1468–1471.
Markey, D. (1986) *On the edge of Empire: foodways in Western Australia, 1829–1979,* unpublished PhD thesis, Pennsylvania State University.
Marsden, D. (1969) *Mothers Alone: Poverty and the Fatherless Family,* London: Allen Lane.
Marshall, L. (1976) *The !Kung of Nyae Nyae,* Cambridge, Massachusetts: Harvard University Press.
Marvin, G. (1988) *Bullfight,* Oxford: Basil Blackwell.
Maslow, A.H. (1943) 'A Theory of Human Motivation', *Psychological Review* 50: 370–396.
Massarik, J. (1987) 'Rusty Bruno cautious over Tyson fight', *Guardian,* 26 March: 28.
Matthews, W. and Wells, D. (1982) *Second Book of Food and Nutrition,* London: Forbes.
Mauss, M. (1925) *The Gift,* 1954 edn, London: Cohen & West.
Meat Trades Journal (1987) 'Meat is just a way of life for British families',

London: International Thomson, 2 July letter: 2.
Meat Trades Journal (1989a) 'Chicken tops new poll', London: International Thomson, 20 July: 15.
—— (1989b) 'Farmers flout loophole in hormone ban', London: International Thomson, 27 July: 1.
—— (1989c) 'Beef suspect source of rare organism', London: International Thomson, 27 July: 15.
—— (1989d) 'Chicken sales "hit more" by poison scare', London: International Thomson, 27 July: 20.
—— (1989e) 'Smile please!', London: International Thomson, 3 August: 1.
—— (1989f) 'Sales slump in wake of poisonings', London: International Thomson, 3 August: 3.
—— (1989g) 'Good welfare saves money', London: International Thomson, 10 August: 20.
Mennell, S. (1985) *All Manners of Food,* Oxford: Blackwell.
Merchant, C. (1982) *The Death of Nature: Women, Ecology and the Scientific Revolution,* London: Wildwood House.
Michell, K. (1987) *Practically Macrobiotic,* Wellingborough: Thorsons.
Midgley, M. (1980) *Beast and Man,* London: Methuen.
Miles, E. (1904) *A Boy's Control and Self Expression,* Cambridge: E. Miles.
Millett, K. (1977) *Sexual Politics,* London: Virago (1969).
Mills, J. (1989) *Womanwords: A Vocabulary of Culture and Patriarchal Society,* Harlow, Essex: Longman.
Mitchison, R. (1985) 'A Case Study to Illustrate how the Demand by a Rich Area can Enable Poorer Areas to Obtain Money for Grain Purchase and thus Reduce Malnutrition', in Centre for Human Ecology (ed.): 3.
Monsanto (1976) 'More Food', company advertisement in *Scientific American* 253, 3: 124.
More, H. (1655) *An Antidote Against Atheism,* 1662 revised edn, Cambridge: Morden.
Moscovici, S. (1976) *Society Against Nature: The Emergence of Human Societies,* London: Harvester.
Murcott, A. (1982) 'On the Social Significance of the "Cooked Dinner" in South Wales', *Social Science Information* 21, 4–5: 677–696.
Naipaul, V.S. (1964) *An Area of Darkness,* London: Andre Deutsch.
Nakornthab, P. (1986) *Urbanization and national developments: A study of Thailand's local urban governments,* unpublished PhD thesis, Cornell University.
Nash, R. (1967) *The Wilderness and the American Mind,* London: Yale University Press.
Needham, R. (ed.) (1973) *Right and Left: Essays on Dual Symbolic Classification,* London: University of Chicago Press.
—— (1983) *Against the Tranquility of Axioms,* London: University of California Press.
New Internationalist (1987), reporting 'World Food Association Bulletin', 3 and 4, 1986, July 1987.
Niven, C. (1967) *History of the Humane Movement,* London: Johnson.

Noren, K. (1987) *Studies on organochlorine contaminants in human milk (pesticides, polychlorinated biphenyls, dioxins),* unpublished MEDDR thesis, Karolinska Institutet, Sweden.

Novak, G. (1971) *Empiricism and its Evolution,* New York: Pathfinder Press.

O'Brien, P.K. (1977) 'Agriculture and the Industrial Revolution', *Economic History Review,* 2nd series XXX: 166–181.

OED (1973) *The Shorter Oxford English Dictionary,* Oxford: Oxford University Press.

Opschoor, H. (1985) 'EC Policies, European Meat Consumption and Environmental Degradation in Botswana', in Centre for Human Ecology (ed.) *Focus on Meat:* 18–19, transcript of proceedings of European Workshop in Human Ecology, 27–31 May, University of Edinburgh.

Ortner, S. (1982) 'Is Female to Male as Nature is to Culture?', in M. Evans (ed.) *The Wamn Question:* 485–507 London: Fontana.

Ortner, S. and Whitehead, H. (eds) (1981) *Sexual Meanings,* London: Cambridge University Press.

Orwell, G. (1937) *The Road to Wigan Pier,* 1984 edn, Harmondsworth: Penguin.

Palin, M., Chapman, G., Cleese, J., Idle, E., Jones, T., and Gilliam, T. (1973) *The Brand New Monty Python Bok,* London: Eyre Methuen.

Parker, A. (1987) 'Till diet do us part', *Sun,* 4 November: 13.

Parrinder, P. (1989) 'Conrad and Eliot and Prejudice', letter to the *London Review of Books,* 14 September 1989: 4.

Pember Reeves, M. (1979) *Round about a Pound a Week,* London: Virago.

Perisse, J., Sizaret, F. and Francois, P. (1969) 'The Effects of Income on the Structure of Diet', *Nutrition Newsletter.*

Perl, L. (1974) *The Hamburger Book,* New York: Seabury.

Perren, R. (1978) *The Meat Trade in Britain 1840–1914,* London: Routledge & Kegan Paul.

Peters, G.H., Jones, G.T., and Hyder, K. (1983) *A Review of the Factors Influencing the Demand for Meats in the United Kingdom,* University of Oxford: Institute of Agricultural Economics.

Pimental, D. and Pimental, M. (1979) *Food, Energy and Society,* London: Edward Arnold.

Polanyi, K. (1977) *The Livelihood of Man,* New York: Academic Press.

Poole, F.J.P. (1983) 'Cannibals, Tricksters and Witches: anthropophagic images among Bimin-Kuskusmin', in D. Tuzin and P. Brown (eds) *The Ethnography of Cannabilism,* Washington: Society for Psychological Anthropology.

Pullar, P. (1970) *Consuming Passions: A History of English Food and Appetite,* London: Hamish Hamilton.

Punch, M. (1977) *Progressive Retreat: A Sociological Study of Dartington Hall School,* Cambridge: Cambridge University Press.

Radcliffe-Brown, A.R. (1922) *The Andaman Islanders,* 1964 edn, New York: Free Press.

Raine, C. (1989) 'Conrad and Prejudice', *London Review of Books,* 22 June 1989: 16–18.

Read, P.P. (1974) *Alive: The Story of the Andes Survivors,* London: Secker & Warburg.

Realeat Surveys (1984–1990), Gallup Surveys into meat-eating and vegetarianism conducted for the Realeat Company Ltd, London.

Reclus, E. (1901) 'On Vegetarianism', *Humane Review* January: 316–324, undated edn, reprinted by Practical Parasite Publications, London.

Reeves, P. (1988) 'Spanish butter wouldn't melt in their mouths', *Independent*, 25 January: 1.

Regan, T. (1985) *The Case for Animal Rights*, in P. Singer (ed.) *In Defence of Animals*: 13–26, Oxford: Blackwell.

Renfrew, J. (1985a) *Food and Cooking in Prehistoric Britain: History and Recipes*, London: English Heritage.

Renfrew, J. (1985b) *Food and Cooking in Roman Britain: History and Recipes*, English Heritage.

Renner, H.D. (1944) *The Origin of Food Habits*, London: Faber & Faber.

Reuter (1990) 'East eats West as Big Mac arrives in Moscow', *Guardian*, 1 Feb: 20.

Richards, A. (1939) *Land, Labour and Diet in Northern Rhodesia*, London: Oxford University Press.

Riches, D. (1982) *Northern Nomadic Hunter-Gatherers*, London: Academic Press.

Rivers, J. (1981) *An Historical Perspective on Nutrition*, London School of Hygiene and Tropical Medicine: MSc & Diploma in Human Nutrition.

Rivière, P.G. (1980) Review of Arens (1979), *Man* 15, 1: 203–205.

Robertson, V. (1989a) 'Let's talk turkey on welfare of livestock', *The Scotsman*, 6 August: 18.

—— (1989b) 'Protests could cook goose for grouse on 12th', *The Scotsman*, 11 August: 3.

Robertson Smith, W. (1889) *The Religion of the Semites*, Edinburgh: Black.

Rombauer, I.S. and Rombauer Becker, M. (1931) *Joy of Cooking*, 1946 edn, London: Dent.

Rosaldo, M. and Atkinson, J. (1975) 'Man the Hunter and Woman', in R. Willis (ed.) *The Interpretation of Symbolism*: 43–75, ASA Studies 2, London: Malaby Press.

Rouse, I.L., Armstrong, B.K. and Beilin, L.J. (1983a) 'The Relationship of Blood Pressure to Diet and Lifestyle in Two Religious Populations', *Journal of Hypertension* 1: 65–71.

Rouse, I.L., Beilin, L.J., Armstrong, B.K., and Vandongen, R. (1983b) 'Blood Pressure Lowering Effect of a Vegetarian Diet: Controlled Trial in Normotensive Subjects', *Lancet* 1: 3–10.

Rowntree, S. (1913) *How the Labourers Live*, London: Nelson.

Rozin, P. (1976) 'Psychobiological and Cultural Determinants of Food-choice', in T. Silverstone (ed.) *Appetite and Food Intake*, Life Sciences Research Report 2, Dahlem Workshop on Appetite and Food Intake: 285–312, Berlin: Dahlem Conferenzen.

Russell, B. (1946) *History of Western Philosophy*, London: Allen & Unwin.

Sacks, F.M., Castelli, W.P., Donner, A. and Kass, E.H. (1975) 'Plasma lipids and lipoproteins in Vegetarians and Controls', *New England Journal of Medicine* 292: 1148–51.

Sahlins, M. (1976) *Culture and Practical Reason*, Chicago: University of Chicago Press.

Sahlins, M. (1983) 'Raw Women, Cooked Men, and Other "Great Things" of the Fiji Islands', in D. Tuzin and P. Brown (eds) *The Ethnography of Cannibalism*: 72–93, Washington: Society for Psychological Anthropology.

Sanday, P.R. (1986) *Divine Hunger: Cannibalism as a Cultural System*, Cambridge: Cambridge University Press.

Schwartz, W. (1989) 'Archbishop Preaches a Green Gospel', *Guardian*, 18 September: 2.

Scott, W. (1829) *The Betrothed*, 1894 edn, London: John Nimmo.

Seeger, A. (1981) *Nature and Society in Central Brazil: The Suya Indians of Mato Grosso*, London: Harvard University Press.

Serpell, J. (1986) *In the Company of Animals*, Oxford: Blackwell.

Sharp, H. (1981) 'The Null Case: The Chipewyan', in F. Dahlberg (ed.) *Woman the Gatherer*: 221–244, New Haven: Yale University Press.

Shelley, P.B. (1813) *The Complete Poetical Works of Shelley*, 1904 edn, T. Hutchinson (ed.), Oxford: Clarendon.

Sherratt, A. (1981) 'Plough and Pastoralism: Aspects of the Secondary Products Revolution', in I. Hodder, G. Isaac, and N. Hammond (eds) *Pattern of the Past: Studies in Honour of David Clarke*: 261–305, Cambridge: Cambridge University Press.

Shields, S. (1986) *An economic history of nineteenth-century Mosul*, unpublished PhD thesis, University of Chicago.

Shoard, M. (1987) *This Land is Our Land*, London: Paladin.

Shostak, M. (1983) *Nisa: The Life and Words of a !Kung Woman*, Harmondsworth: Penguin.

Simmel, G. (1907) *The Philosophy of Money*, 1978 edn, London: Routledge.

Simoons, F. (1967) *Eat Not This Flesh: Food Avoidances in the Old World*, Madison, Wisconsin: University of Wisconsin Press.

Singer, P. (1976) *Animal Liberation: A New Ethics for our Treatment of Animals*, London: Jonathan Cape.

Siskind, J. (1973) *To Hunt in the Morning*, Oxford: Oxford University Press.

Skramlik, van (1926) *Handbuch der Physiologie der Niederen Sinne*, cited in Renner (1944): 19–20.

Slocum, S. (1982) *Woman the Gatherer: Male Bias in Anthropology*, in M. Evans (ed.) *The Woman Question*: 473–484, London: Fontana.

Sloyan, M. (1985) 'Economic Importance of the Meat Trade', in Centre for Human Ecology (ed.) *Focus on Meat*: 3–4, transcript of proceedings of European Workshop in Human Ecology, 27–31 May, University of Edinburgh.

Smith, E. (1863) *Sixth Report of the Medical Officer to the Privy Council*, Appendix No. 6, report by Dr Edward Smith on the Food of the Poorer Labouring Classes in England.

Smith, G. (1988) 'How Lucy lives on £4.35 a week!', *Edinburgh Evening News*, 9 December: 14.

Social Trends 20 (1990) Government Statistical Service, London: HMSO.

Soler, J. (1979) 'The Semiotics of Food in the Bible', in R. Foster and O. Ranum (eds) *Food and Drink in History: Selections from the Annales, Economies, Civilisations*: 126–138, Baltimore: Penguin.

Spencer, C. (1988) 'Kibbutzburgers and spring buds, *The Guardian,* 12 March: 23.
Sperber, D. (1975) *Rethinking Symbolism,* Cambridge: Cambridge University Press.
Spiegel, M. (1988) *The Dreaded Comparison: Human and Animal Slavery,* London: Heretic books.
Spring Rice, M. (1981) *Working-Class Wives,* London: Virago.
Stead, J. (1985) *Food and Cooking in 18th Century Britain: History and Recipes,* London: English Heritage.
Stewart, J. (1989) 'The meat trade needs to change – right now!', *Butcher & Processor,* July: 7.
Stewart, P. (1985) 'European Meat Eating – The World Context', in Centre for Human Ecology (ed.) *Focus on Meat*: 4–6, transcript of proceedings of European Workshop in Human Ecology, 27–31 May, University of Edinburgh.
Strathern, M. (1980) 'No Nature, No Culture: The Hagen Case', in C. MacCormack and M. Strathern (eds) *Nature, Culture and Gender*: 174–222, Cambridge: Cambridge University Press.
Tambiah, S. (1985) *Culture, Thought and Social Action,* Cambridge, Massachusetts: Harvard University Press.
Tannahill, R. (1988) *Food in History,* revised edn, London: Penguin.
Teherani-Kronner, P. (1985) 'An Opportunity for, or an Obstacle to, Ecologically Sound Agriculture – meat production in the FGR and the Liquid Manure Order of North Rhine Westfalia', in Centre for Human Ecology (ed.) *Focus on Meat*: 12–14, transcript of proceedings of European Workshop in Human Ecology, 27–31 May, University of Edinburgh.
Teleki, G. and Harding, R. (1981) *Omnivorous Primates: Gathering and Hunting in Human Evolution,* New York: Colombia University Press.
Teuteberg, H.J. (1986) 'Periods and Turning-points in the History of European Diet: A Preliminary Outline of Problems and Methods', in A. Fenton and E. Kisbán (eds) *Food in Change: Eating Habits from the Middle Ages to the Present Day*: 11–23, Edinburgh: John Donald.
Thirsk, J. (1978) *Horses in Early Modern England,* Reading: Reading University Press.
Thomas, K. (1983) *Man and the Natural World,* Harmondsworth: Penguin.
Todhunter, E.N. (1973) 'Food Habits, Food Faddism and Nutrition', in M. Rechcigl (ed.) *Food, Nutrition and Health: World Review of Nutrition and Dietetics* 16: 186–317, Basel: Karger.
Trueman, I. (1989) 'Jap who ate our troops for dinner'/'Pilot on the menu at Jap general's dinner' *Daily Star,* 2 February: 1, 5.
Tudge, C. (1985) *The Food Connection: The BBC Guide to Healthy Eating,* London: BBC Publications.
Turner, E.S. (1964) *Heaven in a Rage,* London: Michael Joseph.
Tuzin, D. and Brown, P. (eds) (1983) *The Ethnography of Cannibalism,* Washington: Society for Psychological Anthropology.
Twigg, J. (1983) 'Vegetarianism and the Meanings of Meat', in A. Murcott (ed.) *The Sociology of Food and Eating: Essays on the Sociological Significance of Food,* Aldershot: Gower.

Veblen, T. (1899) *The Theory of the Leisure Class*, 1959 edn, London: Allen & Unwin.

Voinovich, V. (1987) 'The bald and the hairy', *The Guardian*, 28 December: 13.

Voltaire (1764) *Philosophical Dictionary*, 1971 edn, Harmondsworth: Penguin.

Vulliamy, E. (1988) 'Fur trade hits back at leg trap labels', *The Guardian*, 7 April: 6.

Wallace, A.F.C. (1956) 'Revitalisation Movements', *American Anthropologist* 58, 2: 264–281.

Washburn, S. and Lancaster, C.S. (1968) 'The Evolution of Hunting', in R. Lee and I. DeVore (eds) *Man the Hunter*: 293–303, Chicago: Aldine.

Weinstein, L. and de-Man, A. (1982) 'Vegetarianism vs. Meatarianism and Emotional Upset', *Bulletin of the Psychonomic Society* 19, 2: 99–100.

Weiskel, T. (1989) 'The Ecological Lessons of the Past? An Anthropology of Environmental Decline', *The Ecologist* 19, 3: 98–103.

Wellenkamp, J. (1984) *A psychological study of loss and death among the Toraja (Sulawesi, Indonesia; funerals)*, unpublished PhD thesis, University of California, San Diego.

Westermarck, E. (1924) *The Origin and Development of the Moral Ideas*, London: Macmillan.

White, L. (1949) 'Energy and the Evolution of Culture', in L. White (ed.) *The Science of Culture*, New York: Farrar, Strauss.

White, L. Jnr. (1967) 'The Historical Roots of our Ecological Crisis', *Science* 155: 1203–1207.

Williams, W. (1977) 'UK Food Production: Resources and Alternatives', *New Scientist* 76, 1081: 626–628.

Wilson, P. and Lawrence, A. (1985) 'Improved Biological and Energetic Efficiencies of Meat versus Crop Production', in Centre for Human Ecology (ed.) *Focus on Meat*: 24–26, proceedings of European Workshop in Human Ecology, 27–31 May, University of Edinburgh.

Wollaston, W. (1722) *The Religion of Nature Delineated*, London: Knapton.

Young, J.Z. (1968) 'Influence of the Mouth on the Evolution of the Brain', in P. Person (ed.) *Biology of the Brain*, Washington: American Association for the Advancement of Science.

Zeuner, F.E. (1963) *A History of Domesticated Animals*, New York: Harper & Row.

Zola, I. (1972) 'Medicine as an Institution of Social Control', *Sociological Review* 20: 487–504.

NAME INDEX

SUBJECT INDEX